24.95

Presented to

Major General R. W. Carpenter

from

Brigadier Bruce Cook ADC
Director General Reserves
Australian Army

ABCA Reserves Conference

September 2009

DEFENDERS OF AUSTRALIA: THE 3RD AUSTRALIAN DIVISION, 1916–1991

THE AUSTRALIAN ARMY HISTORY SERIES

Winning with Intelligence
Judy Thomas

The Warrior Poets
Robert Morrison

The History of the Royal Australian Corps of Transport 1973–2000
Albert Palazzo

DEFENDERS OF AUSTRALIA: THE 3RD AUSTRALIAN DIVISION, 1916–1991

ALBERT PALAZZO

Army History Unit
Department of Defence
Canberra ACT
2002

First published in Australia – 2002
by AUSTRALIAN MILITARY HISTORY PUBLICATIONS
13 Veronica Place, Loftus 2232 Australia
Phone: 02-9542-6771 Fax: 02-9542-6787

Written by Albert Palazzo

Printed in Australia by Ligare Pty. Ltd. – Riverwood NSW.
Typesetting by Margaret McNally
Index by Damien Fenton

National Library of Australia
ISBN: 1-876439-03-3

To the citizen soldiers of the 3rd Division, and their families and communities, who gave so much over 75 years for Australia's sake, in peace and war.

Contents

List of Tables

List of Maps

List of Illustrations

Acknowledgements

As a historian, I am always indebted to numerous individuals and institutions for their support in my research and writing. This project was no exception. Over the course of its preparation, I had the pleasure of meeting and communicating with numerous veterans of the 3rd Australian Division. Invariably, these citizen soldiers gave unselfishly of their time and memories. The response to my appeal was so great, there was not enough room to identify everyone individually in the Division's story. To those who are not mentioned, I offer my apology, but also note that your experiences form part of the fabric of this history.

I owe a special debt of gratitude to several members of the 3rd Division for their unceasing efforts to provide me with the information I sought. These include Saul Spielman and Leon Stubbings of 3rd Division Signals, Ian Rainford of the 58th/32nd Battalion, and Neil Leckie of 22nd Battalion RVR. A special note of thanks is extended to Kevin Cooke, a former commander of the 3rd Division and the moving force behind this project. Kevin continually offered his support and assistance, and I thank him for his patience.

I would also like to thank several associations of the 3rd Division for providing me with access to their members, and for sharing materials on their unit's history. They include the HQ 3rd Australian Division (AIF) Association, Royal Victoria Regiment Association, 3rd Australian Division AASC Association, 58th/59th Battalion Association, 24th Battalion Association, 37th/52nd Australian Infantry Battalion Association, 3rd Division Signals Association, 57th/60th Battalion Association, 58th/32nd Battalion Association, Ex-22nd Battalion RVR Association, and 2nd Field Regiment Association.

I would like to thank my colleagues at the School of History, University of New South Wales at the Australian Defence Force Academy, for providing the cordial atmosphere in which this work was written. The assistance of the Army History Unit was also crucial in the development of this project, from conception to publication. The head of the unit, Roger Lee, has my thanks. The maps were drawn by Keith Mitchell of the Australian National University, who also has my thanks.

Last, I would like once again to thank my wife, Melissa Benyon. During the years spent on this project, she never stopped her encouragement or tired of reading drafts.

List of Abbreviations

AA	Assistant Adjutant
AA(ACT)	Australian Archives – Canberra
AAMC	Australian Army Medical Corps
AAOC	Australian Army Ordnance Corps
AASC	Australian Army Service Corps
AAVC	Australian Army Veterinary Corps
AA(VIC)	Australian Archives – Melbourne
ADFA	Australian Defence Force Academy
AEME	Australian Electrical & Mechanical Engineers
AGPS	Australian Government Printing Service
AIF	Australian Imperial Force
amb	ambulance
AMF	Australian Military Forces
ANGAU	Australian New Guinea Administrative Unit
ARA	Australian Regular Army
a/tk	anti-tank
AWM	Australian War Memorial
bde	brigade
bn	battalion
bty	battery
CGS	Chief of the General Staff
CMF	Citizen Military Forces
CO	Commanding Officer
coy	company
CPP	Commonwealth Parliamentary Papers
CRA	Commander Royal Artillery
CRAASC	Commander Royal Australian Army Service Corps
CRAEME	Commander Royal Australian Electrical & Mechanical Engineers
det	detachment
div.	division
E & ME	Electrical and Mechanical Engineers
eng.	engineers
fd	field
GOC	General Officer Commanding

gp	group
HMT	His Majesty's Transport
HQ	Headquarters
hyg	hygiene
inf	infantry
int.	intelligence
LAD	Light Aid Detachment
MBI	Military Board Instruction
OC	Officer Commanding
pl	platoon
PMF	Permanent Military Forces
QMG	Quarter Master-General
RAA	Royal Australian Artillery
RAAF	Royal Australian Air Force
RAASC	Royal Australian Army Service Corps
RAE	Royal Australian Engineers
RAEME	Royal Australian Electrical & Mechanical Engineers
RAR	Royal Australian Regiment
recce	reconnaissance
regt	regiment
RQR	Royal Queensland Regiment
RSAR	Royal South Australia Regiment
RTR	Royal Tasmania Regiment
RVR	Royal Victoria Regiment
RWAR	Royal Western Australia Regiment
sec.	section
sigs	signals
UK	United Kingdom

Chapter 1
The Origins of the 3rd Australian Division

Introduction

The story of the 3rd Australian Division is, in many ways, the history of a nation's progress to military maturity. The 3rd Division is the longest serving division in the Australian Army. It was raised on 2 February 1916 during World War I and remained on the order of battle until its disbandment on 29 June 1991.

The only gaps in these 75 years of service were two brief periods after the world wars, when the division was in hiatus while the army's senior officers determined the force's post-war organisation. In each case, the army soon revived the 3rd Division, although other divisions were not so fortunate.

The Commonwealth formed the 3rd Division early in the nation's life, when Australians' consciousness of themselves as a separate people had just begun to emerge on the battlefields of Gallipoli and the Western Front. During the inter-war period, the 3rd Division was a part of Australia's defence preparations, and would have been at the forefront of the struggle had Japan invaded. During World War II the Division served overseas, playing a major role in halting the Japanese thrust in New Guinea and re-conquering Australia's mandate territories. After World War II the Division's task was to defend Australia from the communist threat during the Cold War.

The 3rd Division's place in the military history of Australia is much greater, however, than preparation for and participation in war. Of equal importance was its role in the development of the Australian Army. As the force's longest serving division, it played a part in most of the events, decisions and policies that determined the development of the Australian Army, and helped to develop the nature of service in the nation's military forces. As a middle level headquarters, the 3rd Division was rarely in a position to initiate army policy but did help shape it and put it into effect. The history of the Division and of the Australian Army are intertwined and one cannot be told fully without the other. The 3rd Division's history is thus an effective tool with which to examine the broad factors that have shaped Australian military experience, in both peace and war.

A number of major trends guided the development of the Australian Army and the 3rd Division. They include the army's tradition of citizen soldiery, the rivalry which existed between the regular and militia branches of the force, the limitations on overseas service imposed by the *Defence Act*, the army's relationship with government, and the need for the force to change in response to the evolving nature of warfare. How each of these factors affected the 3rd Division was critical in its development. To different degrees each provided certain benefits to the division. However, each, again to different degrees, also contained liabilities. In the end, these factors also provided the army with a rationale for its disbandment.

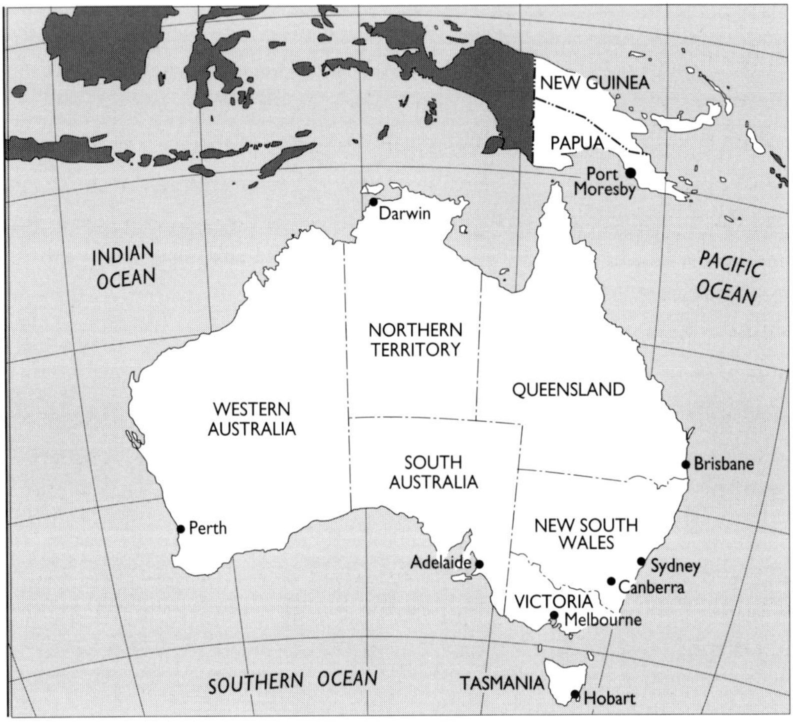

MAP 1. Australian Territory

One of the most enduring elements which shaped the Division is the incorporation of Australia's citizen soldier tradition into its ethos. The 3rd Division always comprised citizen soldiers, representing a tradition of the citizen soldier as the state's primary defender. Citizen soldiers were concurrently members of the army and the community. These men (and later women) were part-time warriors, who gave military service in addition to the full-time duties they performed in their communities. One of the designations given to this type of service conveys the sense that service in the division was a community duty: the Citizen Military Force or CMF. The 3rd Division was one of the key formations in the CMF until, during its final years, the army replaced citizen soldiery with the current army reserve system.

Australia's regular soldiers too are drawn from the community and return to it upon discharge. However, important differences exist. The regulars are full-time soldiers and do not hold other positions or occupations in society. They are not permanent members of the local community, as the army periodically relocates them to other barracks. Although members of the 3rd Division served full-time during both world wars, they were not the equivalent of regulars. There was never any question of the Division's wartime personnel remaining in permanent military service when peace returned. Demobilisation saw them rapidly return to civilian roles. Their martial obligation retreated again to the background, to await their nation's next call.

The Australian citizen soldier tradition arose in the colonial period, as part of the volunteer soldier movement which flourished throughout the British Empire during the 19th century.[1] Militia units existed in Australia as early as 1792, when the commandant of Norfolk Island organised free settlers into a local body. Most early citizen soldier units had only a brief existence as enthusiasm tended to wane as soon as the immediate threat which had caused their raising was past. For example, after an initial influx of recruits to counter the Russian threat in 1854, during the Crimean War, most volunteer units had disappeared by the conflict's end. However, after Britain withdrew its garrison regiments from Australia in 1870, the volunteer movement gained greater permanency. In 1901, on the eve of Federation, the forces of the Australian colonial armies numbered some 27,000 citizen soldiers.[2]

The Division's citizen soldier tradition has had both positive and negative effects on its development. One benefit was the close links between units and the community, which helped recruitment and continuity of membership, and gave soldiers a highly personal reason for service. However, the notion that citizen soldiers were 'natural' soldiers has been exaggerated, largely by C. E. W. Bean, who raised the idea of the instinctive military effectiveness of the World War I digger to mythical heights. Citizen soldiers are not 'natural warriors' and the lesson of the world wars was that they required the same degree of training as the regular soldier. After World War II, when the government raised an infantry brigade of full-time regular soldiers, the competence of the CMF soldiers underwent greater scrutiny. The unsubstantiated claim of CMF superiority provided the regular forces with ammunition that contributed to the CMF demise.

Although the citizen soldier formed the basis of Australian defence during the colonial era and for much of the Federation period, the Commonwealth has nearly as long a tradition of employing long-service, full-time soldiers. In 1870, after Britain withdrew its imperial garrision, colonial governments gradually established a second military organisation to serve beside its citizen units. In 1871, New South Wales raised Australia's first paid, full-time and permanent military force. The other colonies soon followed. However, the colonial governments did not intend that the permanent soldiers should become the primary means of territorial defence, and did not raise permanent combat units. Instead of infantry battalions or cavalry regiments, the full-time personnel served as staff officers, training instructors and gunnery and engineering troops, staffing the colonies' military headquarters and providing administrators and instructors to the CMF. They also garrisoned the continent's coastal defences, maintaining them in a state of readiness. Until recently, these permanent soldiers represented only a tiny proportion of

Australia's military strength, and most troops served in the CMF. At the time of Federation, the colonies had a total of only about 1500 permanent troops on their establishments.[3]

During Federation, the regular soldiers became known as the Permanent Military Force, or PMF, in order to distinguish them from the CMF. The *Commonwealth Defence Act* of 1903 recognised the division in the Army's organisation and formally split the military into permanent and citizen branches. The decision to maintain those permanent troops marked the beginning of an internal struggle to determine whether the CMF or the PMF would control the army. For much of the 3rd Division's history, governments clearly recognised the CMF as the dominant branch. However, in the 1950s, government attitudes began to change, and control within Army Headquarters shifted decisively in favour of the PMF.

The establishment of this two-army system was a major factor in the history of the 3rd Division. The regulars naturally formed their own caste and assumed that as professionals they were superior to the more numerous, but less highly trained, citizen soldiers.[4] This attitude helped to foster a schism within the army that divided the force into two intellectual and cultural parts. While not particularly marked at the nation's beginning, a rivalry developed between the CMF and PMF during the inter-war period. Tensions continued to simmer during World War II, although the struggle against Germany and Japan kept things from escalating further. However, after the war, the government expanded the PMF and raised the nation's first non-citizen soldier infantry battalions. From that point, the rivalry between the two halves of the force became more intense, and the regulars used their position to undermine the privileges of the CMF. The contest ended in the dismemberment of the CMF and its replacement by the Regular Army Reserve. The disbandment of the 3rd Division arose directly from the CMF's failure to win the army's internal struggle.[5]

Although they were members of the British Empire, Australia's soldiers were not under any obligation to serve outside the borders of their colonies or, later, those of the federated nation. Colonial and Federation governments resisted frequent entreaties from the authorities in London to place Australian forces under British command as part of an Imperial army. For this reason, the *Defence Act* forbade the government to despatch any of its citizen soldiers abroad without first obtaining their individual permissions. The government could compel its forces to defend only the physical territory of their colony or that of the nation. If the government wanted to despatch military forces outside the country, it had to form special units composed of volunteers.

The statutory basis for this prohibition was the *Commonwealth Defence Act*, which stated, "Members of the Military Forces shall not be required, unless they voluntarily agree to do so, to serve beyond the limits of the Commonwealth and those of any Territory of the Commonwealth".[6] During the Cold War, and especially after Australia's recent intervention in East Timor, the government made some modifications to the overseas service clause of the *Defence Act*, making it far easier to employ the army's reserve forces (the modern equivalent of the CMF) outside national territory. Yet even today, some limitations remain on the use of reserve soldiers.

Australians have never been reluctant soldiers. They willingly came forward when called upon by the Empire. Their first overseas imperial expedition was the Maori Wars in New Zealand. In 1885, the British expedition to the Sudan, to avenge the death of General Charles Gordon, attracted a contingent of volunteers from New South Wales. At the time of Federation, large numbers of Australians served in colonial, and later Commonwealth, units in South Africa during the war against the Boers. Australians also assisted in the relief of Peking during the Boxer Rebellion of 1901. When the government formed the 1st and 2nd AIFs respectively for service in Europe, all members were volunteers.

There is one small exception to the government's observance of the limitation on overseas service. During the darkest days of World War II, the government of John Curtin legislated a compulsory overseas service obligation which extended the citizen soldier's territorial service obligation to include most of the South West Pacific Area, a zone which incorporated the Dutch East Indies, the Philippines, New Guinea and the Solomon Islands. For the first time, the government could compel CMF personnel to serve beyond Australian territory. However, the Curtin government barely used its new powers and the change proved insignificant. CMF units, including the 3rd Division, fought the Japanese in Papua and on New Britain, Bougainville and adjacent islands, but these were Australian territory, held either as colonies or as League of Nations mandates. The sole CMF force to serve outside Commonwealth territory was the 11th Brigade, which garrisoned Merauke in Dutch New Guinea – still within 80 kilometres of the border with Australian Papua. When the conflict ended the government restored the pre-war conditions of service.

The legislative limitation on overseas service defined the nature of citizen soldier service in Australia. The CMF existed exclusively for the defence of Australian territory, and was never meant to act as an expeditionary force. The intention was to prevent the Australian government from being able to despatch troops at will on Imperial expeditions, or the incorporation of Australian troops into a standing Imperial army under the control of the War Office in London. The *Defence Act* ensured government moderation in military adventure and the Commonwealth's military forces remained truly Australia's Army.

However, as the nature of war changed, the prohibition on overseas service transformed the citizen soldier's position in the defence hierarchy from one of strength to one of weakness. During the Cold War, the government wanted troops not only for mainland defence but for contributions to Allied efforts in a number of small anti-communist wars in Asia. The Korean War, the Malayan Emergency, Confrontation and the Vietnam War were all fought by the regular Army. Changing defence priorities at that time made it difficult for the CMF to maintain its status as the nation's primary defender, especially when the regular Army was defending Australia from offshore. The limitation on voluntary service gradually helped to consolidate the growing ascendancy of the regular soldier, and contributed to the demise of the CMF and its 3rd Division.

Australia's citizen soldiers have not been well served by the nation's political leaders. At best, the government's relationship with its armed forces, both the citizen and permanent elements, can be described as benign neglect. Only recently did the nation's armed forces become the Commonwealth's first line of defence. From Federation until after World War II, the Australian government did not consider defence as one of its duties as a sovereign state. Instead, Australia's primary defender was the Royal Navy. Imperial authorities, and a succession of Australian governments, believed that the Royal Navy would either deter an enemy from attacking the Australian mainland or defeat anyone foolhardy enough to make the attempt. Even as the power of the Royal Navy declined during the inter-war period, the Australian government still held to the promise of the infallibility of the fleet. Only after the fall of Singapore did the government recognise the necessity to provide for Australia's defence from its own resources. Then Australia turned to another benefactor – the United States. After World War II, Australia's alliance with the United States became the cornerstone of its defence policy.

The consequences of depending on a 'big brother' were serious. Because Britain and then the United States had undertaken to protect Australia, the Commonwealth government funded its forces so inadequately that in some years the lack of money threatened the very survival of the army as a functioning organisation. The army often had to curtail training programs, defer the purchase of modern equipment, and maintain extremely low stockpiles of critical items. While 'big brother' was available, the government neither had to confront the realities of security planning nor address the needs of its armed forces.

As one of the major CMF formations, the 3rd Division was greatly affected by the government's neglect. Members of the Division have had to make do with broken or obsolete equipment, and limited, unimaginative training opportunities. Derelict drill halls, badly designed uniforms, low pay and poor benefits provided little incentive for recruits or serving personnel. Fiscal priorities, not national security issues, were always the primary factor in the government's decision-making. In fact, the Commonwealth has routinely made major defence-related policy decisions solely on the basis of their effect on the budget. The decision to abolish the CMF, and to disband the 3rd Division were, in part, motivated by fiscal considerations.

Over the 3rd Division's long history, change was to be expected. The Division Australia raised in 1916 was not the same as that which the army disbanded in 1991. How the Division responded to change forms an important part of its story. Even something as fundamental as the definition of what constituted a citizen soldier did not remain constant. For example, the members of the division, while part-time soldiers (except during the world wars), have not always been volunteers. At various times, the government has authorised conscription. While an individual could always volunteer for service in the CMF, during part of the inter-war period, and two episodes during the Cold War era, the government imposed a mandatory service obligation on the nation's young men. However, the citizen soldier tradition remained intact, and conscription did not make recruits into regular soldiers. Instead, they continued to serve on a limited basis and retained their obligations to their local communities.

The ability to modify its organisation and capability to meet change posed one of the greatest challenges to the leaders of the 3rd Division. In particular, the Division found it difficult to respond to the policy changes and developments in the nature of modern warfare which appeared during the latter part of the Cold War era. For example, the government's decision to pursue a policy of forward defence to the north increased its reliance on regular soldiers, producing a gradual shift of resources away from the CMF to the regulars. In addition, the decision to move the army's heartland from Victoria and New South Wales to the nation's far north meant that most citizen soldier units lost touch with the force's centre. The primary problem the CMF faced in the second half of the Cold War lay in defining for itself a mission central to the government's defence policy. Its failure to do so brought about its end as a force in the military hierarchy, and contributed to the disbandment of the 3rd Division.

For many years, the 3rd Division was a formation on the Australian Army's order of battle, with a distinguished history and lengthy and successful service in two world wars. The periods of greatest difficulty for the division occurred not in those relatively few years of war but in the much longer periods of peace. Then, the government's interest in military matters, and especially the CMF, declined, and the 3rd Division fared poorly in the struggle for resources and status. How the Division managed during these peacetime battles, and how it coped with the various changes that shaped the development of the Australian Army, were as decisive in determining the Division's history as its experience in wartime.

The Decision to Raise the 3rd Division

In 1910, the government of Andrew Fisher accepted the proposals of Field Marshal Lord Kitchener for the organisation of the army. As an Imperial officer, Kitchener had toured Australia to advise on the Commonwealth's defence requirements. He recommended a military organisation of 84 battalions of infantry, organised into 21 brigades. He also called for 28 regiments of Light Horse, 56 batteries of guns and howitzers and fourteen field companies of engineers. The government subsequently raised the Army's size to 92 battalions and 23 brigades. Kitchener did not see a need for any divisions since he assumed that the Australians would serve within British divisions.

Service in the force was by two methods. Men could either join the CMF as volunteers or serve under the obligations of the government's universal service program. This conscription scheme required all eligible males between the years of eighteen and 26 to train for sixteen days per year, including an eight-day camp. Members of the artillery and engineers had to train for 25 days, of which seventeen were in camp. The only exception was in the soldier's final year of obligation, when he had just one muster parade.[7]

While it was not fully implemented by 1914, the army had made good progress towards creating the Kitchener organisation. By the end of fiscal year 1913/14, it had formed 52 battalions of infantry, 23 regiments of light horse and 27 batteries of field artillery. The force included 12,314 volunteers and 33,601 conscripts. The outbreak of war interrupted the further development of the force.[8]

There was never any doubt that Australia would come to Britain's aid when war broke out. When news of Britain's declaration of war against Germany reached Australia, the Prime Minister, Joseph Cook, proclaimed, 'When the Empire is at war, so is Australia at war'. The then leader of the opposition, Andrew Fisher added that Australia would stand besides the mother country to 'our last man and our last shilling'.[9] The first shot fired by the British Empire in World War I was actually in Melbourne. The gunners of Fort Nepean, at the mouth of Port Phillip, directed a warning shot across the bow of the German merchantman *Pflaz*, which was attempting to reach the high seas. Under the watchful gaze of the fort's guns, the ship turned back and returned to Port Melbourne.

The immediate dilemma the government faced was the need to organise an expeditionary force to assist Britain against Germany. As the *Defence Act* prohibited the despatch of troops outside the nation's territory, the government could not simply mobilise the units of its existing army. Instead it had to create a new force composed of soldiers who had volunteered to serve overseas. This was the origin of the 1st AIF.

The first voluntarily enlisted unit was the Australian Naval and Military Expeditionary Force which, after a hasty raising, sailed on 19 August to capture the German colonial territory of New Guinea. With the surrender of the German Governor at Rabaul on 13 September, its mission was completed. Australia's main contribution, however, was the AIF, which was raised in a more considered manner. The government had agreed to provide Britain with 20,000 men, organised into an infantry division and a light horse brigade. The task to raise and command these troops fell to Major-General William T. Bridges, the Army's Inspector-General.

Bridges decided to create the AIF on a territorial basis. Each State would contribute troops according to its population. For example, New South Wales provided an infantry brigade, a field artillery brigade and a regiment of light horse, whereas Tasmania despatched half a battalion of infantry, a field battery and a squadron of light horse. These were the basis of the 1st Australian Infantry Division. The response to the call was so great, however, that Australia was able to offer Britain an additional infantry brigade and two further brigades of mounted troops.[10]

The AIF was soon en route to Egypt, where it began training and awaited a decision as to its destination. The Anglo-Franco plan to force the straits linking the Mediterranean and Black Seas determined that the Diggers' first battle would be at Gallipoli instead of on the Western Front. At first light on 25 April 1915, the 1st Division came ashore at ANZAC Cove.

After the campaign's failure, and its evacuation from Gallipoli in December 1915, the AIF returned to Egypt for a period of rest and reorganisation, and a decision on its future employment. After deliberating with Britain, the Commonwealth government agreed that most of the AIF should move to France, to fight the Germans on the Western Front. Only the Light Horse was to remain in the desert for use against the Ottoman Empire.

1: Members of 11th Machine Gun Company at camp in Seymour, Victoria.
[*Courtesy* AWM P02057.001]

The Australian government also agreed to a massive expansion of the AIF. At that time, the Australian contingent consisted of two divisions, the 1st and 2nd, and two independent brigades, the 4th and 8th, as well as light horse units. In Egypt, the AIF also had a large pool of infantry reinforcements awaiting assignment to a unit. On 2 February 1916, the Australian government cabled London, accepting Britain's suggestion to expand the AIF to five divisions. As a result, the AIF reorganised its forces in Egypt to create two new divisions, while the Commonwealth raised a fifth at home. Those formed in Egypt became known as the 4th and 5th Divisions. Although the one raised in Australia was the last to arrive in France, the army designated it as the 3rd.[11]

To its sister divisions, the members of the 3rd became known as the 'hard thinkers' or the 'procrastinators', nicknames which came about because its members had not enlisted in the first romantic rush to the colours after the war's commencement. Instead, by the time they joined, the Great War had been under way for a year and a half. Australia's losses had already reached more than 8400 killed and about 20,000 wounded.[12] Although the 3rd had delayed joining, the course of the war had only steeled their determination to do their part. Unlike those of the 1st Division, the men who joined the 3rd had no doubt of the risky course upon which they embarked.

The older divisions were thus veterans long before their junior sibling took up its place in France, where, unlike the other Australian divisions, it did not participate in the Battle of the Somme. Instead, while the rest of the AIF fought savage battles at Pozières and Mouquet Farm, the 3rd Division trained safely in England, arriving at the front after the Somme Campaign had come to an end. The 3rd Division would have to earn its place in the AIF.

One of the typical 'hard thinkers' was Lieutenant C. H. Alexander of the 9th Light Trench Mortar Battery. He waited until the government had made a special appeal before volunteering in mid-1915, when he believed that it was no longer right to hold back. Alexander knew what to expect, and hoped that he would not see any fighting, but had also recognised that the time had come to put everything else aside until the war was won.[13]

While perhaps 'hard thinkers', men still joined the 3rd Division for myriad reasons. For some, there was still an element of romance in enlisting. One new Digger in the 37th Battalion remembered the party-like enthusiasm of his mates as their transport pulled away from the wharf in Melbourne on its voyage to England. J. S. Bambrick wrote that the 'boys are dead excited ... all rushing around and singing "I'm going to leave the old home" [and] "Goodbye girlee"'.[14] Another young recruit first tried to join the AIF at the age of sixteen and was initially successful but, after his father's intervention, was sent home. Nothing could keep him out of the war, and the next year his father allowed him to try again. He successfully enlisted in the 38th Battalion, then forming up at Bendigo.[15]

For others, the primary rationale was loyalty. The mother country needed everyone's help and there was a vital job to be done. Colin H. Butler joined the 41st Battalion because he believed that, to achieve his ambition to be a sheep station owner, he had to do his part in helping defeat Germany. He perceived his future in terms of the preservation of the British Empire. G. P. Cuttriss shared Butler's beliefs, enlisting because he believed it was necessary to crush the spirit of militarism which dominated Germany. He hoped that a new Germany would emerge in the war's aftermath.[16] C. H. Linklater, a captain in the 33rd Battalion, wrote his 'fighting blood has been awakened now at seeing what Britain is doing and I will do my utmost to keep the flag flying'.[17]

As the casualties mounted, enlistment conditions had also become less restrictive, giving previously rejected men a second chance. By July 1915, the Army had lowered the minimum height requirement for recruits to just five feet two inches. V. G. Schwinghammer first attempted to enlist during the initial rush to colours, but failed his medical examination. When the Army raised the 3rd Division, he tried again and was soon a private in the 42nd Battalion. Schwinghammer wrote that, this time, the Medical Officer gave his approval without bothering to look too closely. Cyril Herborn, who joined the 33rd Battalion, had a similar experience. Rejected as medically unfit after the war's outbreak, he had no difficulty enlisting in early 1916. Others refused to let rejection on medical grounds dissuade them from their goal to serve Australia. Thomas Kaighin attempted to join up in 1915 but failed because of his teeth. He sought out a dentist, had the problem fixed and, in 1916, tried again. This time he passed and was soon a member of the 30th Battery.[18]

The men of the 3rd Division had little doubt about their status as soldiers. They perceived themselves neither as regulars nor as possessors of any particularly martial spirit. An account of the 39th Battalion, after its initiation into battle at Messines in June 1917, stated that the unit was composed of 'farmers, clerks, vendors, etc.'. It continued that the men of the unit could take pride in their accomplishments for, although new to soldiering, they had been able to take on Germans who had been 'preparing for 40 years'. The members of the 3rd would become efficient soldiers, but they always remained citizens in uniform.[19] Another nickname bestowed on the 3rd Division was 'eggs-a-cook'. The expression came from the streets of Cairo, where the veterans of Gallipoli bought hard boiled eggs from stalls. The vendors would call out 'eggs-a-cook, verra s[w]eet, verra clean'. When 3rd Division troops appeared in France, their seniors in the self-styled 'fighting divisions' would yell out, 'Here they are. Eggs-a-cook. Verra nice, verra sweet, verra clean' – to suggest that the inexperienced new arrivals were about as capable of fighting as a boiled egg. That the 3rd Division's colour patch had an oval, egg-like shape did not help. Invariably, the result of these taunts was a brawl. However, it did not take the 3rd Division very long to earn the respect of the rest of the AIF. Within a short time the 'eggs-a-cooks' could stand face to face with the enemy and make their own contribution to the growing military reputation of the Australian soldier.[20]

Endnotes

1 A full discussion of the citizen soldier movement in Australia is found in Craig Wilcox, *For Hearths and Homes: Citizen Soldiering in Australia, 1854-1945* (St Leonards: Allen & Unwin, 1998); and Bob Nicholls, *The Colonial Volunteers: The Defence Forces of the Australian Colonies, 1836-1901* (North Sydney: Allen & Unwin, 1988).

2 Figures taken from 'Report on the Department of Defence for the Period from 1st March 1901 to 30th June 1906', *CPP*, vol. 2, 1906.

3 Jeffrey Grey, *A Military History of Australia* (Melbourne: Cambridge University Press, 1999), p. 40. Figures taken from 'Report on the Department of Defence for the Period from 1st March 1901 to 30th June 1906', *CPP*, vol. 2, 1906.

4 Peter Dennis et al, *The Oxford Companion to Australian Military History* (Melbourne: Oxford University Press, 1997), p. 163.

5 A good example of the rivalry between the CMF and PMF is found in David Horner, 'Staff Corps Versus Militia: The Australian Experience in World War II', *Defence Force Journal*, no. 26 (Jan/Feb 1981), pp. 13-25.

6 Quoted in Ian D. Rae, 'History of the Australian Army, Part 1: Early History', pp. 13-15.

7 'A Brief Narrative of the Organization, Strength and Training of the Australian Military Forces Since Federation', Military Board Proceedings, AA(ACT), A2653/1, 1928, vol. 1, folder 18; and 'Defence of Australia, Memorandum by Field Marshal Viscount Kitchener of Khartoum', in *CPP*, vol. 2, 1910. An outline of the army's pre-Kitchener organisation is found in 'Defence Scheme for the Commonwealth of Australia', AWM113, item MH1/3.

8 Ian Hamilton, *Report on an Inspection of the Military Forces on the Commonwealth of Australia* (Melbourne: Government Printer, 1914), p. 14; and G. M. Kirkpatrick, *Annual Report of the Inspector-General of the Military Forces* (Melbourne: Government Printer, 1913).

9 Quoted in Ernest Scott, *Australia During the War* (Sydney: Angus and Robertson, 1937), p. 22.

10 'Bridges to Commandants of Military Districts', 10 August 1914, AA(ACT), A2657/T1, vol. 2.

11 'The Formation of the 3rd Division, AIF In 1916 and Service in World War I' (author's possession).

12 Figures from A. G. Butler, *The Australian Army Medical Service in the War of 1914-18*, vol. III, *Special Problems and Services* (Canberra: Australian War Memorial, 1943), pp. 895, 900.

13 Bill Gammage, *The Broken Years: Australian Soldiers in the Great War* (Canberra: Australian National University Press, 1974), p. 14.

14 *Ibid.*, p. 29.

15 J. N. I. Dawes and L. L. Robson, *Citizen to Soldier: Australia Before the Great War, Recollections of Members of the First AIF* (Carlton, Vic: Melbourne University Press, 1977), p. 175.

16 *Ibid.*, pp. 142-43, 175; and G. P. Cuttriss, *'Over the Top' With the Third Australian Division* (London: Charles H. Kelly, 1918), p. 78.

17 Quoted in Gammage, *The Broken Years*, p. 42.

18 Diary, Schwinghammer Collection, AWM, 2DRL234; John Edwards, *Never a Backward Step: A History of the First 33rd Battalion, AIF* (Grafton: Bettong Books, 1996), p. 22; and Thomas R. L. Kaighin, *A Personal Account of World War I*, transcribed by Rodd Johnson (Rodd Johnson, 2000).

19 'Our Stunt, An Account of the 39th Battalion in Messines', 7 June 1917, AWM27, item 111/17.

20 *42nd Battalion, AIF* (Brisbane, 1938), pp. 15-16.

Chapter 2
The 3rd Australian Division in World War I

Forming the Division

During February and into March 1916, the Commonwealth assembled the 3rd Division. Following existing practice, recruitment was on a regional basis, each State contributing a proportion of the formation's arms. The 9th Brigade, containing the 33rd–36th Infantry Battalions, came solely from New South Wales. The 10th Brigade contained three Victorian battalions – the 37th, 38th and 39th – and the 40th Battalion from Tasmania. Queensland provided half of the 11th Brigade – the 41st and 42nd Battalions – while South Australia contributed the 43rd Battalion and Western Australia filled out the formation with the 44th Battalion.

The other arms were raised on a similar basis. For example, although the 3rd Pioneer Battalion's training centre was at Campbellfield, Victoria, only its Headquarters and A Company came from that State. Its B Company was from New South Wales, its C Company originated in Queensland, and Western Australia and South Australia each provided half of its D Company. The Army raised the 11th Field Ambulance at Mitcham, South Australia, while the 9th Field Ambulance came together at Liverpool Camp, New South Wales.[1]

The training the troops received while in Australia was little more than basic drill, physical exercise and adjustment to army life. During the 3rd Pioneer Battalion's time at Campbellfield the men gained little experience in their primary duty – the constructing of trench works – owing to the shortage of engineering stores, and their skills did not advance beyond the level of elementary attack formations. The 38th Battalion's period at Bendigo was similar. When the time came for its embarkation, the men could march and drill but had little training in fighting. As the unit's historian noted, the men had the appearance of trained soldiers but not the skills. After the Division's artillery brigades arrived in England their commander, Brigadier-General H. W. Grimwade, observed that they had had little practice with guns and were virtually untrained.[2]

The 43rd Battalion did its initial training at Morphettville, in Adelaide. Most of the work consisted of parades, drill and route marches. The troops did gain a modicum of musketry experience, at a miniature range on the dry bed of the Sturt River and at an open range in the sandhills at Glenelg and Henley. The battalion's most complex exercise occurred when it undertook a mock battle, attacking the Hindmarsh Bridge via Montefiore

Hill. While the troops were enthusiastic, the value of the training was doubtful. There was no effort to systematise the various units' programs into a single and coherent syllabus, and some of the instruction the men received was outdated. For example, the 37th Battalion practised an attack in which its companies opened out from column formation into extended order, before advancing towards their objectives. On the Western Front, such a manoeuvre would have been suicidal. Serious training had to wait until the Division arrived in England. There, the British had constructed suitable facilities, staffed by instructors experienced on the Western Front.[3]

2: Members of the 8th Field Artillery Brigade ready to sail from Port Melbourne on HMT *Medic*, May 1916. [*Courtesy* AWM PB0571]

During May and June, depending on the availability of shipping, the Division's units broke camp, proceeded to harbours and embarked on transports for the long voyage to England. When the 43rd Battalion's turn came on 9 June crowds lined the route and filled the quay alongside the *Afric* to bid the men farewell. Also on board was the 25th Company ASC and part of the 11th Field Company. On the *Runic*, so many men climbed into the ship's rigging to wave goodbye that the vessel looked as though it had been draped in khaki. Others unrolled hundreds of paper streamers to friends below, connecting the *Runic* to the pier with a maze of coloured ribbon. When the 44th Battalion sailed, its unit historian described the wharf as a blaze of colour, with flags, handkerchiefs and streamers filling the view. Jim Roberts sailed on the *Persic* with the 37th Battalion. His mates crammed the railing of the ship, waving to friends below while a band played 'Australia Will Be There', 'Goodbye Girlie', and 'Tipperary'. Roberts managed to pick out his cousins in the crowd and, catching a line they threw him, hauled up a supply of chocolate and some messages. Others around him, he noted, were doing the same. As the ships slowly pushed off from the wharf, the streamers which had symbolically connected the men to Australia parted one by one and well-wishers gathered them up to keep as mementoes.[4]

Six months later, the ritual of departure had changed little. Private V. G. Schwinghammer sailed on 16 November 1916 on the *Kyarra*, as part of the fifth reinforcements of the 42nd Battalion. Bands and well-wishers again packed the wharf, while soldiers played out the paper streamers to their friends below. Schwinghammer sailed with his brother Charlie, who was destined for the 41st Battalion.[5]

For most, the trip to England was uneventful, even monotonous. The experience of the 3rd Pioneer Battalion was fairly typical. It sailed from Port Melbourne on the *Wandilla* on 6 June. The first stop was Perth, where the ship coaled before heading across the Indian Ocean, to pause briefly at Durban and then sail on to Capetown. It then headed north into the Atlantic Ocean, re-coaling at Cape Verde Islands. Picking up a destroyer escort off the coast of Spain, the *Wandilla* arrived at Devonport on 26 July. While most transports went via Cape Town, a few took the Mediterranean route. After leaving Sydney on 2 May, the 34th Battalion stopped at Albany, Colombo, Suez, Port Said, and Alexandria, before reaching Devonport on 23 June. The *Afric*, carrying the 43rd Battalion, also went via Egypt but, instead of going all the way by sea, disembarked the unit at Marseilles. It crossed France by rail. At Le Havre, the unit transferred to channel steamers and arrived at Southampton on the 24 July, thus becoming the first unit of the 3rd Division to reach France, if only in transit.[6]

The level of accommodation on the transports was, of course, a function of rank. Captain W. J. Rose of the Pioneers had a berth cabin to himself. He wrote home to say that the *Wandilla* was exceedingly comfortable and that his meals, staterooms and steward service were of first class passenger level.[7] Other ranks lived below the officer cabins on troop decks which had been constructed from former hold space. The quarters provided for the 39th Battalion were standard. During conversion to a transport, yard workers had installed a series of tables on each troop deck. Each table was known as a mess and had an allotment of twenty men. Overhead were hooks, from which the men hung their hammocks in the evening, while racks lined the wall for stowing equipment. Each mess appointed two orderlies from its number to carry the table's food from the galley and do the cleaning up. In return, the orderlies were exempted from other shipboard duties.[8]

The universal affliction on leaving the port was seasickness. Few men had much experience at sea travel and, once the transports began to plough through the broad rollers of the Southern Ocean, only a handful avoided illness. Rose had to abstain from all meals for the first day, while Private S. M. Bellingham in the 38th Battalion took a few days to gather his sea legs. Roberts admitted to having given the fish 'a good blow-out'.[9] More serious was the threat of disease. Inevitably, on the crowded vessels, men became ill and the threat of an epidemic was real. Most units suffered their first active service casualties during the voyage. The *Runic* arrived at Cape Town with eighteen cases of measles, including two fatalities, while on board the *Kyarra* meningitis claimed three lives.[10]

There was little to do during the lengthy voyage. The officers alleviated boredom and kept the men out of mischief by organising activities to keep everyone busy. Boat drills were held regularly and Bellingham wrote of the need to report to his station with his lifebelt and water bottle. Sport was popular. Bellingham believed that his outfit – D company – was going to win the tug of war, as it had already bested the engineers of 10th Field Company, who shared the *Runic* with the 38th Battalion. In the Pioneers, boxing matches caused much interest. Lectures on the war, tactics or ports of call also provided entertainment. Some formations managed to smuggle a mascot aboard: 10th Field Company had a kangaroo. Frequent deck concerts put on by unit bands also helped pass the time. On some transports, the men provided their own entertainment. On arriving at Fremantle the men on the *Kyarra* put up two shillings a piece and went into town to buy a piano. The ship's captain permitted them to strap it down on the main top deck, where it served as a centre of activity for the rest of the voyage. Schwinghammer, who had some talent at the keyboard, remembered playing every evening for 'buck dances' and sing-songs. He also noted that, on Saturday nights, the men would hold a formal concert, printing programs on the ship's press.[11]

The highlight of each voyage was when the ship crossed the equator. Time-honoured maritime tradition required that anyone crossing the line for the first time had to be presented to King Neptune. All personnel on board were his subjects and the dignity of rank did not protect anyone from appearing before the sea monarch. In anticipation of the event, the *Runic*s crew rigged a large canvas bath on the main deck. The activities began during the night, when the sea king's emissaries roused Captain C. H. Peters of the 38th Battalion from his bed and plunged him and the other first-time officers of the unit into the bath, pyjamas and all. The next afternoon, Peters participated in the formal ceremony, playing the part of Neptune's Queen in the trial of the other initiate officers. Each was brought before the King and charged with offences such as 'being a general nuisance' or 'paying too much attention to dress'. Their sentences were either the administration of a bitter medicine poured down their throats or being lathered with a pink paste and then shaved with an oversize wooden razor. Once the officer's ceremony was over, it was the other ranks' turn, each man being presented to Neptune and then suitably dunked in the bath. On the *Suevic*, the festivities were enjoyed so much that, as the fun was at its height, the tank burst.[12]

Private Cyril Herborn of the 33rd Battalion related a bit of excitement that occurred on the transport *Port Sydney* as it crossed the Bay of Biscay. An unlucky soldier in the battalion had been enjoying a quiet rest in a deck chair. The sea was rough and suddenly a particularly large swell broke over the side, lifted the chair, and its inmate, and tossed them overboard. A quick-thinking mate threw a life buoy, to which the soldier clung, in freezing water, for two hours, while the ship's captain stopped the vessel and organised a rescue. The soldier, the ship and indeed the entire ship's company had luck on their side that day because the Bay of Biscay was a prime hunting ground for enemy U-Boats. The *Port Sydney*'s captain, having weighed the risks of stopping against those of a possible revolt on board, thought it best to retrieve the water-logged soldier, although the safety of the vessel could have justified leaving him behind.[13]

One of the things the troops most looked forward to was the opportunity to explore a strange port. Unfortunately, as most ships had some ill men on board, the port authorities routinely quarantined the vessels. Oddly, this designation was selective and did not prevent the men from going to shore, merely requiring them to be escorted. Instead of enjoying the freedom of Cape Town, for example, the officers marched the men into the hills overlooking Table Bay. One exception was the *Runic*, whose passengers were allowed shore leave, but, as Bellingham observed, the men could only go ashore if escorted by a non-commissioned officer during the day or an officer at night. He was probably astute in deducing that the restrictions were a result of Diggers on previous transports 'playing up' in town. This helps to explain the strange quarantine regulation that allowed the men ashore for route marches but not for visits to the town's drinking establishments.[14]

Some soldiers were determined to enjoy the delights of a foreign port at all costs. While the 33rd Battalion was on a route march in the hills above Cape Town, Corporal L. R. Boulton noticed that a lot of the men had broken away and cleared out. That night, he observed still more going over the side of the ship and making their way into town. Boulton was on guard duty that evening and had to go into town with a patrol and round them up. Lieutenant E. M. Brissenden of 3rd Division Headquarters recorded that, during his stay in Cape Town, the troops were not allowed ashore. However, the guard was insufficient and 120 men made their escape from the ship in two massed rushes, although three men were bayoneted. A bold and ingenious member of the 39th Battalion made it ashore by climbing into a basket being used to coal the *Ascanius*. By the time the basket had been lifted back onto the wharf, he was covered with a fine black powder so dense that he was able to leave the dockyard with a party of African stevedores. Others, it appears, overdid their ill-gotten liberty. When the *Ascanius* sailed from Capetown it left two of its company behind; while the *Persic* left four. These men followed on later transports.[15]

As the ships approached their destination, the danger of German submarines increased. When any vessel reached the latitude of Cape Verde Islands or entered the Mediterranean Sea – depending on the route – the captain instituted special precautions. On the *Kyarra*, the crew darkened the portholes and the ship travelled without lights at night. To help protect the ship, soldiers armed with rifles did duty as submarine lookouts, the officers placing them around the ship and instructing them to fire at any U-boat's periscope they sighted. Port authorities at Cape Town and Port Said also mounted 4.7 inch naval guns on each vessel's deck to enable a transport to fight off a submarine. As the 3rd Division approached England, destroyers met the transports and shepherded them to port.

Sometimes the escorts failed to rendez-vous. When this happened the transport's captain had no choice but to proceed. The men on the *Runic* spent two nervous moonlit nights before safely making port.[16]

Upon arrival, the troops disembarked and transferred by rail to the training establishment on Salisbury Plain. The experience of Sapper Spencer E. Jenkins of the 3rd Division Signal Company was typical. His transport docked at Plymouth. After a few hours' leave in the port area, where Jenkins got to explore a battleship, his unit entrained, and they proceeded to Exeter across the Devon countryside, which Jenkins found beautiful. Arriving at 1 a.m., the train paused briefly while the Mayoress personally distributed refreshments, as she did for all troop trains which passed through her city. The train continued through Salisbury to Amesbury, where the troops detrained at 5 a.m. The signallers then had a five-mile march to Lark Hill, the 3rd Division's home for the next few months.[17]

Lark Hill was only a small part of the British Army's vast training establishment on Salisbury Plain. Here, for the first time, members of the Division met their commander, Major-General John Monash, a highly regarded officer with considerable combat experience gained at Gallipoli and on the Western Front.

Monash had a clear idea of what he wanted to accomplish at Salisbury. He intended to train the Division vigorously so that, when his troops entered the trenches for the first time, they would have the knowledge and skills to survive and to deal the enemy heavy blows. Monash was not an innovative leader, but did believe in learning from the experience of others. He had studied the lessons of the Western Front and made sure they were disseminated through his command.[18]

Monash also planned that the Division should have a distinctive identity which, he believed, would further instil an *esprit de corps* throughout the formation. There, he made one of his few, although minor, errors in appraising the spirit of the men. He insisted that they wear their hats brim down, unlike the rest of the AIF, who looped theirs upward. To the relief of the men, General William Birdwood reversed this order upon the formation of the Australian Corps.

Monash was clearly pleased with the potential of his new command and he observed:

> I am bound to say … that the raw material of this division is far and away better than any raw material that we have previously had to handle … There is a certain air about the men which was absent on previous occasions. There are very few young men, most of them averaging about 25 years of age and they all have a mature, independent, hard, active look, the outstanding characteristic being intelligence.[19]

While the natural quality of the men provided a sound foundation for teaching them their trade, Monash was well aware that they still had a long way to go to be considered effective soldiers. One problem was that the Division was still a collection of individuals rather than an organised fighting machine. Before being concentrated at Lark Hill, none of the brigades had met in one place with all their sub-units. The men had no rifles or bayonets – firearms were to be issued in England – and had never seen a Mills Bomb,

Vickers or Lewis gun, dug a trench or handled transport. The artillery was no better since the gunners, while in Australia, had never worked the 18 pounder gun nor the 4.5 inch howitzer. One of the few areas of determined instruction undertaken in Australia had been bayonet training. However, what the men had received was outdated and they had to re-learn close combat fighting. One commentator noted that essentially all the men could do was stand in straight lines.[20]

Monash outlined a thirteen-week training program, of which four weeks would be spent training at the company level, four at battalion level, three at brigade level and, finally, two weeks at divisional level. He also took advantage of the British training system and sent hundreds of specialists off to schools. Each unit thus had a core of trained men who could serve as instructors for the remainder. This allowed everyone in the unit to gain rapid operational familiarity with all the weapons at their disposal. Monash also required that the units undergo the same exercises, down to platoon level, to instil tactical uniformity throughout the Division.

One limitation he faced, however, was a shortage of equipment. Although British factories were producing war materials at levels unimagined before the start of the conflict, demand continued to outstrip supply. The 3rd Division had to wait until production had satisfied higher priorities. Persistent equipment shortages, Monash believed, accounted for the delay in the formation's transfer to France. Guns for the artillery were particularly scarce and, even late in the training program, the field batteries had not received their full allocation of weapons. To alleviate the problem, the Division's artillery and engineering units did some training at British Army depots, where they had access to the full range of weaponry.[21]

To condition the men for France, Monash required units to undertake a program of field exercises, culminating in mock operations in a system of realistic trenches the Division had constructed nearby. The program began with simple night marches and field bivouacs and soon advanced to night patrolling, sniping, fighting while wearing a gas mask, dugout construction and other essential tasks. On one occasion when the 39th Battalion was training, it rained in torrents and the men had to find shelter wherever they could, in cow sheds and haystacks – certainly a foretaste of what to expect on the Western Front. Despite Monash's insistence on realism, however, it was still only practice. Sometimes officialdom interfered with training. While Bellingham was training in a network of poorly maintained trenches near Lark Hill, a medical officer appeared and condemned the site as unfit for troops, summarily terminating the exercise. Bellingham acknowledged that the conditions were pretty rough, but pointed out things were no easier in France.[22]

The 3rd Division's most advanced training ground was the trenches at Bustard, six miles from Lark Hill. There, the Division undertook brigade-level exercises. The 10th Brigade had a dreary time there during its rotation. It was brutally cold, and incessant rain transformed the trenches into swollen streams. Verey lights illuminated a bleak landscape, while rifle cracks and grenade explosions broke the silence. The men spent the night in sodden overcoats, awaiting the dawn. The brigade kept two battalions in the front line, with the other two in support, and mock raids and night patrols kept everyone interested. On one occasion, Captain A. Tootell exploded a mine comprising 5000 pounds of gun cotton. The resulting crater was 30 feet deep and 80 feet across. The waiting infantry then charged the position and established a defensive strong point incorporating the crater.[23]

Monash's methods brought results. In late August, the British Army's Inspector of Infantry, Major-General Francis Howard presented to Monash a generally favourable report on that arm's progress. Despite continuing deficiencies in some types of equipment – for example, the battalions still only had 150 rifles each and no telescopic sights – the men were no longer the raw recruits who had left Australia a few months earlier. Howard remarked, 'The division is going to come on well, it is being trained on sound lines'. He observed that the men were of fine physique, were keen, and were developing quickly.[24]

However, in the midst of its preparations, the Division was almost disbanded. The other four Australian divisions had suffered heavy casualties during the Battle of the Somme and, as the rate of enlistment at home had declined, it was unclear how the Commonwealth would bring them back up to strength. On 18 August, the War Office asked Monash how many of his men were immediately available for despatch to France, and how many would be ready in three months' time. Monash realised the seriousness of the request. He replied that the Division's only fully trained men were battalion or company commanders, who had been specifically allocated to him because their experience facilitated the training of the rest of the formation. He emphasised that Australia had raised the Division from untrained men, that the instruction program he had prepared aimed to create a uniformly trained formation, not a succession of replacement drafts. He lost the debate, however. Ten days later, he had to agree to the immediate transfer of 2800 men to other units, 233 from each battalion, although no further drafts proved necessary.

Later in the year the Australian Government reopened the question of the Division's survival. In the wake of the rejection of conscription by Australian voters, the Australian Prime Minister, William Hughes, suggested that the AIF should use the Division as a depot to maintain the other formations. The War Office rejected that advice and ordered the Division to France; far from disbanding them, London asked the Australian Government to maintain its units at full strength.[25]

On 27 September, King George V inspected the full Division at Bulford, a short march from Lark Hill. It was a major event, and 20,000 men passed their monarch in review. Lieutenant Clive Crowley, of the 33rd Battalion, thought the march past presented the Division in a fine light. The battalions lined up in platoon formation – 22 men abreast with twelve paces between platoons – took well over an hour to go by. Gunner L. W. Robinson marched close enough to the King to hear him comment to an aide, 'A very fine line'. Afterwards, the men lined the road along which the King departed and cheered him as he rode past. Crowley later heard that the demonstration had impressed the King and that Monash thought the formation was too good to remain much longer in Salisbury. Crowley echoed the sentiment, writing that he did not want to wait much longer before moving to France.[26]

The men of the 3rd Division did not have much longer to wait. In mid-November they began to move across the Channel.

Table 1 outlines the Division's organisation for the coming battle on the Western Front.

The 3rd Division at War, 1916–17

By the time the 3rd Division arrived in France in late 1916, planning for the British Army's 1917 campaign was already well under way. Despite its training at Lark Hill, the Division was inexperienced. When the troops entered the trenches for the first time, they would have two objectives; to learn how to fight and survive on the Western Front, and to play their part in the army's offensive program. General Douglas Haig, Commander of the British Expeditionary Force believed in a policy of relentlessly 'wearing down' the enemy. His objective was that his forces should gradually develop a superiority over the enemy in morale and skill at arms, while at the same time reducing German manpower and exhausting the enemy's reserves. His objective was not attrition for its own sake. Rather, he was convinced that the wearing down of the enemy, and the establishment of British superiority, was an essential prerequisite for victory, and that this victory could only occur after his forces had bested the Germans in a decisive battle. He envisaged the battle as of such scale and intensity that at its end the enemy would be prostrate, unable to resist the British (and Allied) forces further. 'Wearing down' and the decisive battle were therefore linked and together formed the operational objectives of the British Expeditionary Force.

In December 1916, Haig had entered into an agreement with General Robert Nivelle, the new French Commander-in-Chief on the Western Front, that the French Army would strike the decisive blow in April of the following year. Haig accepted that his force's role was to undertake a series of diversionary attacks in order to make the Germans commit their reserves prematurely. The agreement was to lead on 9 April 1917 to the battles of Arras and Vimy. In January 1917, however, Haig modified the agreement with Nivelle by extracting a promise that, should the French offensive not succeed, the British could stop their diversionary attacks and redirect their own efforts further north, to Flanders. On 16 April, the French duly launched a war-winning offensive on the Aisne, but instead of victory, reaped defeat – their worst in the war. The French Army then mutinied and Britain's ally did not play a major offensive role on the Western Front again until late in the war. Haig was now free to make his own attempt at decisive victory. The 3rd Division would participate in Haig's bid for victory in the Passchendaele campaign, as well as the 7 June preliminary attack on Messines.

Meanwhile, Haig did not expect his army to spend the winter of 1916–17 in idleness. He instructed his subordinates to undertake minor operations, to use up German manpower and attain superiority over the enemy. Once the 3rd Division moved into the line, it also would participate in this policy. The 3rd Division thus spent its first winter in France aggressively contesting the control of no man's land and raiding German lines.

Table 1
Organisation of the 3rd Division
on the Western Front[27]

CORPS	COMPONENTS
Headquarters	3rd Division Headquarters
Artillery	Headquarters 3rd Division Artillery
	7th Field Artillery Brigade HQ
	• 25th, 26th, 27th Batteries
	• 107th Howitzer Battery
	8th Field Artillery Brigade HQ
	• 29th, 30th, 31st Batteries
	• 108th Howitzer Battery
	3rd Division Ammunition Column
	5th Australian Medium Trench Mortar Battery
	6th Australian Medium Trench Mortar Battery
Engineers	Headquarters 3rd Division Engineers
	9th, 10th, 11th Field Companies
	3rd Division Signal Company
Infantry	Headquarters 9th Infantry Brigade
	33rd Bn The New England Regiment
	34th Bn The Illawarra Regiment
	35th Bn Newcastle's Own Regiment
	36th Bn St George's English Rifle Regiment
	9th Australian Light Trench Mortar Battery
	Headquarters 10th Infantry Brigade
	37th Bn The Henty Regt
	38th Bn The Northern Victoria Regiment
	39th Bn The Hawthorn-Kew Regiment
	40th Bn The Derwent Regiment
	10th Australian Light Trench Mortar Battery
	Headquarters 11th Infantry Brigade
	41st Bn The Byron Scottish
	42nd Bn The Capricornia Regiment
	43rd Bn The Hindmarsh Regiment
	44th Bn The West Australian Rifles
	11th Australian Light Trench Mortar Battery
	3rd Australian Division Pioneer Battalion
Machine Gun	3rd Australian Division Machine Gun Battalion
Transport	22nd, 23rd, 24th, 25th Companies AASC
Medical	9th, 10th, 11th Field Ambulances
	No. 3 Sanitary Section
Veterinary	3rd Division Mobile Veterinary Section
Miscellaneous	3rd Division Salvage Section
	3rd Division Employment Company
	3rd Division Field Bakery
	3rd Division Field Butchery

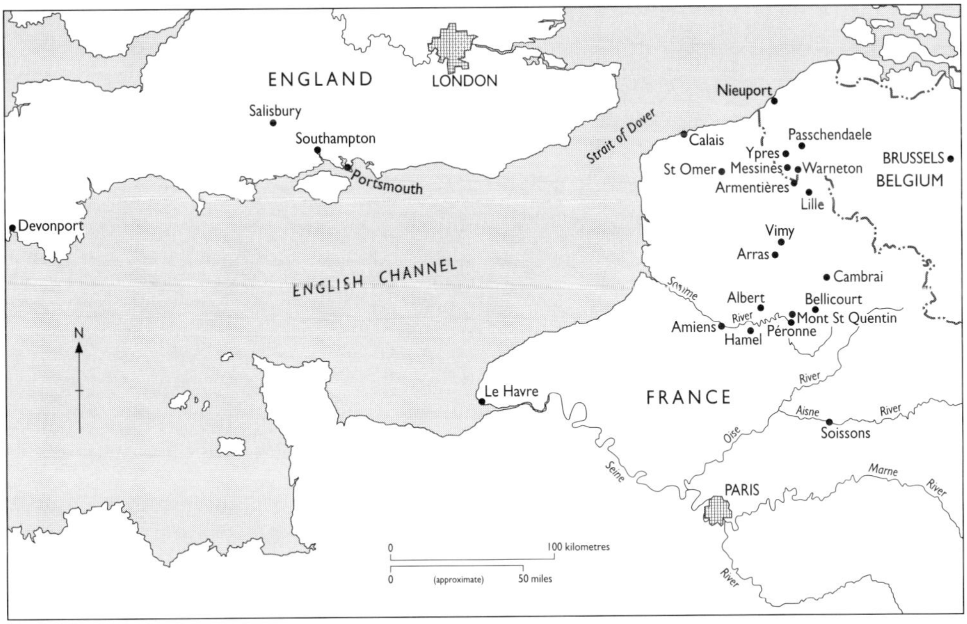

MAP 2. The 3rd Division in World War I

In November 1916, the Division began the long-awaited transfer to France, embarking at Southampton. Each unit made an overnight crossing of the Channel, disembarking at Le Havre on the French side. Unlike the spacious transports employed to bring the men to England, the shuttle across the Channel was in small steamers and paddlewheelers which the port authorities crammed with men and equipment. Jim Roberts, a scout in the 37th Battalion, observed that on the *Lydia* there was room to turn, but not to turn around. His vessel slipped out of the harbour after dark, with the band playing 'Australia Will Be There'. Although tired, he arrived safely the next morning and marched into France. Not everyone had such a smooth passage, however, and when the Channel was rough, seasickness was endemic. The 40th Battalion had a particularly bad crossing, with virtually the entire unit ill.[28]

Each unit's first day on the continent was similar. After the Channel crossing, the troops marched through the city to a rest camp on the outskirts. Unfortunately, their destination, Rest Camp No. 1, was grossly misnamed. Although only seven miles from the wharf, it lay at the top of a hill, and the road up was a long, tortuous climb. According to the historians of the 38th and 40th Battalions, the men, labouring under the burden of their equipment, experienced one of their hardest marches ever. Some 39th Battalion soldiers could not sustain the pace and had to drop out along the route, while in the 37th Battalion, the officers called a rest every half hour. The amount of equipment the men carried was certainly a consideration. One man in the 39th weighed himself on a scale at the dock and discovered that his load totalled 122 pounds. Roberts noted that a young lad in his unit carried a load seven pounds more than he weighed. Not surprisingly, the men arrived at their destination exhausted, fit only for food and sleep. G. P. Cuttriss observed that his icy, rain-swept stay at the camp was one of his most miserable memories of the war. Finally, next morning, the men retraced their steps down the hill to the train station, for transfer by rail to the front.[29]

The division's assembly point was Bailleul, in northern France near the Belgian border. There, the men underwent their final training before moving into the trenches for the first time. The journey from Le Havre was slow and tedious. Packed with all their equipment into France's famous 40 *Hommes* (men) or eight *Chevaux* (horse) trucks, the troops had a cold and uncomfortable time.

For much of the trip, the trains moved so slowly that the men could get out, walk alongside and practise their phrasebook French on the locals. At other times, the train would mysteriously stop for hours until, for equally obscure reasons, the conductor would blow the whistle and everyone would have to leap aboard as the train resumed its torpid crawl. The 9th Field Ambulance's trip, not an uncommon example, took 44 hours to go some 175 kilometres. In one battalion, the story spread that, alighting for one of the train's frequent pauses, a Digger expressed a desire to pick some flowers. His mates pointed out that, as it was only November it was unlikely he would be successful. He replied, 'That's alright, I've got a packet of seeds in my pocket'.[30]

On de-training at Bailleul, the lucky units found buses or lorries waiting to convey them to their billets in nearby villages. Most, however, had to walk the final miles and missed out on the opportunity to enjoy what Australian soldiers derisively referred to as a 'Tommy route march'.[31] The 39th and 40th Battalions settled into Merris, while the 37th and 38th Battalions and the 10th Field Company occupied Strezaele and the 44th Battalion and 11th Field Ambulance moved into Steenwerck. Although these villages were relatively safe, it was clear that the Division's journey was nearly at an end. Its camps were close enough to the front for the troops to hear the sounds of guns firing to the east and to observe aerial contests between hostile planes overhead. At night, the flashes of a barrage illuminated the horizon. Over the next few days, the men received final instructions and stores. The most important were gumboots, essential to protect the feet in the rain-sodden trenches, and the Small Box Respirator, which provided vital protection from the enemy's poisonous gases.

General Headquarters had placed the 3rd Division in Lieutenant-General A. J. Godley's II Anzac Corps, itself a part of General Herbert Plumer's Second Army. Its first place in the line would be the Armentières sector, deliberately chosen as both the British and Germans considered it a quiet sector, and a convenient place to introduce new formations. At Armentières, the men of the 3rd Division began to learn the skills with which to defeat the enemy. These would enable them to assume an equal place of honour with Australia's other divisions. Their stay in that sector demonstrated a mastery of the hazards of trench warfare arising from a carefully calculated progression of exposure to an increasing level of danger.

By the time the 3rd Division arrived at the front, long experience had demonstrated that, before taking over a position, units needed to send forward an advance party. Roberts, a battalion scout, went up the line with a group of other specialists including bombers, snipers and Lewis gunners, to familiarise himself with the intended position, under the guidance of the current occupants, a New Zealand battalion. They spent three nights under the Kiwis' care, being shown the sniping, bombing and Lewis Gun posts, dugouts and store places, and the gaps in the wire. At night, the scouts crept out into no man's land to explore. The New Zealanders also taught the newcomers the hazards of trench life. During dinner, a New Zealander heard a whizzing sound in the air. He told the Australians to hug the side of the trench and a few seconds later, a *minenwerfer* bomb exploded less than 30 yards away. Earth and shell fragments rained down but no one was injured.[32] Once they became experienced, the 3rd Division's reliefs only took an evening.

On 26 November, the 9th Brigade began the process of taking up position in the line. In the evening, advance parties of the 33rd and 34th Battalions made their way forward to the trenches held by the 26th and 27th Northumberland Fusiliers. After dark on the 27th, two sections per platoon relieved half the Fusiliers. The next night, the remaining Australian troops came forward and more of the English soldiers left. The position remained under the command of the British until the 29th, however, at which time the Australians became solely responsible for the sector. While the infantry were exchanged, the 9th Brigade Headquarters followed, taking over from the Headquarters of the 103rd Brigade, and a section of signallers from the 3rd Australian Signal Company replaced its British counterparts.[33]

Over the following weeks, other battalions rotated through a series of assignments, either in the trenches, the support line a little further back or the reserve. The 38th Battalion's initiation was typical. On 1 December, it moved into the line by Houplines, relieving a New Zealand unit. Before entering the communication trench which led forward to their position, the men put on gumboots whose inadequacies soon became apparent. Along the bottom of the communication trench lay duckboards – wooden lattices meant to prevent sinking into the mud – however, many of the duckboards were broken or slippery with slime and a mis-step resulted in immersion into waist-deep mud. C and D Companies had a particularly difficult time forcing their way through a poorly maintained trench called Irish Avenue. Filled with icy water, it resembled a narrow canal rather than a communication line. Arriving at their positions, soaked to the skin and plastered with mud, the men moved into their dugouts. These were little more than hovels carved out of the trench wall and lined with sheets of iron, keeping out neither rain, enemy shells nor rats. For ten days, the men of 38th Battalion stayed there, in conditions their historian could only describe as 'appalling', until the 40th Battalion relieved them on 10 December.[34]

While the infantry acclimatised in the trenches, the other units of the division also got to work. The 11th Field Ambulance assembled in Steenwerck and took over responsibility for a divisional rest station. Soon, its members were at work at medical posts and advanced dressing stations in Armentières. The 9th Field Ambulance arrived at Steenwerck on 27

3: The 42nd Battalion comes out of the line near Houplines during its first winter in France, December 1916. [*Courtesy* AWM E00107]

November and immediately established a field hospital to provide medical facilities for the rest of the Division. The next day, the stretcher bearers in A section took over a regimental aid post and began to bring in the wounded. Meanwhile, C section moved to Pont Nieppe, where it converted the treatment vats of a jute factory into a divisional bath. The engineers and pioneers were also soon at work, repairing dugouts, revetting the sides of communication trenches and digging emplacements for the Division's mortars. In a major project, the 10th Field Company rebuilt the notorious Irish Avenue, raising the duckboards by two feet and increasing the height of the parapet, to improve its drainage and safety.[35]

Despite their preparations in the mock trenches at Bustard, being in the actual lines required adjustment on the part of the men. For Private Bellingham, it was the weather, particularly the cold and rain, that struck him. In February, he noted that the ice was a foot thick. The cold was also a problem for E. A. C. Atkinson of the 9th Machine Gun Company, but it was the rats which moved him. He wrote that there were millions of the beasts and they made life very disagreeable. He also complained that his dugout was so small, he had to sleep with his knees pressed against his head. However, within a short while, he had grown used to the deprivations of trench life. By the following month, he could calmly write home, secure in his dugout, that, despite the bombs bursting around him, he felt quite safe, and if it wasn't for the cold, he would have enjoyed the work.[36]

Although Armentières had a reputation as a quiet sector, it did not remain so for long. The new arrivals immediately began to contest no man's land and engage the enemy. On 29 November, having finalised its relief, the 9th Trench Mortar Battery bombarded the enemy's front and support lines, while the 33rd and 34th Battalions sent out patrols into no man's land after dark. The enemy's mortars replied in kind. Throughout the night, German machine gun fire played along the Australian parapet. By the day's end, the 33rd Battalion had claimed the Division's first victim when one of its snipers potted an enemy soldier. However, the 34th Battalion also reported two casualties.[37]

During his battalion's first night in the trenches, Roberts led a patrol into no man's land. The mission had two objectives: patrol members were to familiarise themselves with the ground between their position and that of the enemy, and seek out German patrols or working parties. It was also an opportunity to practise moving silently over strange terrain and identifying stationary objects in the dark. Roberts and his patrol were out for three hours, sliding in and out of muddy shell-holes and standing or lying still when the Germans illuminated the ground with a flare or search light. On that occasion they did not see any Germans and had only gone halfway, about 100 yards, towards the enemy's defences, but the learning process had begun.[38]

In those early days, even men who did not leave the trenches could find some excitement. One dark night, during the 39th Battalion's first tour in the front lines, a sentry thought he saw something creeping closer. Suddenly, he heard a noise and a black shape loomed before him. The sentry sounded the alarm and, while his mates poured out of their dugouts to man the firestep, he lunged with his bayonet. The squeal of the rat he wounded put an end to the evening's activities.[39]

It was the Germans, however, who initiated the 3rd Division's first combat. On the night of 9 December, a sentry in B Company, 38th Battalion, became suspicious that the enemy might be cutting the wire in front of his position. A Verey light revealed German soldiers in no man's land, seeking cover in shell holes. As the Australians began to fire, the enemy unleashed a trench mortar barrage on their position. Three minutes later, the bombs suddenly stopped and a party of Germans who had lain hidden in front of the Australian position rushed the parapet, throwing grenades into the trench. Amid a shower of bombs, Lieutenant Keith Garland directed the defence of his section, while Private J. J. Meyerlink mounted the parapet and played his Lewis gun on the enemy until the gun malfunctioned. Attempting to repair the weapon, he fell wounded, as did Garland. The Australians suffered four casualties but had achieved their first victory.[40] Three nights later, the Germans tried again, this time against a section of line held by the 39th Battalion. The Germans silently approached the Australian trench and lobbed bombs. The Australians fired their SOS signal, tossed bombs of their own and opened fire with Lewis guns. A minute after the signal's despatch, the 3rd Division's guns responded and pounded no man's land with shells. Once again, the Diggers had repulsed the enemy.[41]

Two weeks later, the 33rd Battalion undertook the Division's first planned infantry attack. On Christmas Eve, a party of Diggers crept through no man's land and gained the

4: Men of the 39th Battalion enjoy a hot meal, December 1916.
[*Courtesy* AWM E00085]

German trenches, but found them deserted. Another operation the following night saw the same result. The 39th and 44th Battalions' first raids were similar. Raiders from the 39th Battalion followed Lieutenant P. L. Smith into the German front line on the night of 9 January. Although they did not see any enemy soldiers, they blew up dugouts, destroyed a machine gun and brought back some captured documents. On 10 January, Captain Charles Biles led a party of 30 men from the 44th Battalion across to the enemy's lines. The men gained the opposing trench and bombed their way in either direction for 50 yards without encountering any enemy. They then withdrew, taking a German sign with them as evidence that they had reached their objective. On the way back, however, they were caught in a German barrage and suffered six wounded. In the latter operation signallers had accompanied the raiders, playing out a field telephone line as they advanced. Twice, enemy shells cut the line and the signallers had to trace the break and repair the connection. Although the raids did not result in any known enemy casualties, they did achieve an objective. The men were learning to fight and now knew that what they could accomplish once, they could do again.[42]

Through the winter, the 3rd Division amassed experience. By the end of their stay at Armentières, C. E. W. Bean believed that the Division had become more proficient at raiding than the senior Australian divisions.[43] The Australians also used the time at

DIEU ET MON DROIT

No 1 BILLET — 3rd Div. Sig. Coy.

IN THE FIELD

CHRISTMAS 1916

DINER.

MENU

POTAGES

Consommé d'Australie — Potage purée de Céleri

PIÈCES

Rosbif — Gigot rôti

LÉGUMES

Pommes de terre bouillies et rôties — Carottes

ENTRÉES

Poulet en Mayonnaise — Poulet rôti — Lapin cuit au four

FROID

Jambon — Sardines — Bœuf pressé

SALADES

Salade d'endive — Salade de Céleri

DESSERT

Compote de Pêches — Plum Pudding au Rhum — Gelée de Groseilles

SAVOURIES

Œufs à la neige — Crême fouettée

BOISSONS

Café — Café au lait
Chocolat — Thé
Vin Rouge — Vin Blanc.

CIGARETTES & CIGARES

5: 3rd Division Signal Company Christmas Dinner Menu, 1916.
[*Courtesy* Alan Jenkins]

Armentières to practice brigade-level attacks. To support these operations Plumer, the Second Army's commander, had formed a mobile force of heavy guns that travelled up and down the front, providing additional fire support for special operations. The troops called this formation the 'Travelling Circus'. On 27 February, when the heavy guns were on hand, the 10th Brigade undertook a major raid, an action that required the co-ordination of all the units under its command. The brigade formed a composite battalion of over 800 officers and men drawn from the 37th and 38th Battalions, as well as specialists from other units such as sappers with demolition training from the 10th Field Company. The 3rd Division's gunners and mortar men were to support the attack by cutting the enemy's wire and, once the assault began, laying a box barrage around the perimeter of the attack zone to trap its defenders and cut them off from reinforcements. The target was a section of line held by the 23rd Bavarian Infantry Regiment.

The troops spent the weeks preceding the raid practising their tasks on a full scale mock-up of the enemy lines. The brigade had accurately replicated the German defences by using information gleaned by aerial reconnaissance photographs. Each raider had a particular place and role in the assault. For example, some men were to rush forward with light bridges to place across the ten-foot-wide barbed wire filled ditch which lay just before the Bavarian position. Others received special instruction in bombing, salvage, wire cutting and stretcher bearing duties. Roberts and the scouts had a particularly important responsibility. To ensure a clear lane through the enemy's wire, his team was to crawl forward during the barrage, place Bangalore torpedoes under the entanglements and detonate them at zero hour. Because only the scouts had the necessary knowledge of no man's land, they had received a task normally the responsibility of engineers. In the dark, with shells exploding on the enemy's parapet, Roberts and his team, Privates C. C. Clayton and F. Davis, and other parties of scouts made their way to the enemy's wire. There Roberts and Clayton placed the torpedo under the entanglements and fell back to a safe distance, playing out a fuse behind them. At the moment the barrage lifted – 12.30 a.m. – they lit the fuse. A few seconds later, the torpedo exploded and the rest of the assaulting troops were up and running towards the enemy trench. The platoon assigned to Roberts' gap brought with it a light bridge to cross the ditch. Then they leapt into the enemy's trench, firing at retreating Bavarians, while parties spread out to throw bombs into dugouts, gather up materials and cause as much havoc as possible. However, Roberts' military career was at an end. While clearing out an enemy dugout, he was wounded in the leg and sent home.

Other parts of the assault did not go as smoothly however, and the raiders gained the enemy's trench only with great determination and at high cost. Finding the way barred by wire, Lieutenant J. W. McDonald (37th Battalion), set his platoon to laying mats to create a path over the obstacle. The Bavarians spotted them and fired a green flare. Soon shells fell among McDonald's party, killing or wounding nearly everyone. McDonald tried again 20 yards to the right, but an enemy machine gun opened fire and scythed through his men. Gathering more troops, he attempted another entry further to the right, and this time succeeded in forcing his way into the enemy's trench. McDonald then made contact with other groups of raiders, while his party proceeded to search and bomb the dugouts in the vicinity. However, lack of time and men prevented him from fulfilling his objective of pressing the attack to the Bavarian second and third lines. Some thirty-five minutes after the raid began, the recall sounded and the Australians streamed back over no man's land, shepherding seventeen prisoners and as much booty as they could carry. The raiders estimated that they had killed at least 200 enemy. The Australian casualties were also high, however, with 32 dead and 103 wounded.[26]

On 13 March, the 'Travelling Circus' was again on the 3rd Division's sector, and the 11th Brigade undertook a big raid. Unfortunately, it had rained heavily during the preceding days and no man's land was a bog. Some officers suggested postponing the operation but, as the heavy guns were available for only a limited period, the attack had to go ahead. Predictably, the results were tragic.

At the commencement of the British barrage, more than 400 men moved off towards the German lines. While they were still struggling through the mud, the enemy artillery started a counter-barrage on no man's land, catching them mired in the open. The Australians pressed forward, but those who had managed to negotiate the defender's wire now confronted another obstacle. Immediately before the German front line lay a deep, water-filled ditch. Plunging forward, they held their rifles and Lewis guns above their heads, at the mercy of the German grenades. Despite these difficulties, a few parties managed to gain the German trench but the enemy soon evicted them and they fell back to their original positions, together with the rest of the force. For the rest of the night, patrols stole out to no man's land to bring in the wounded. The attack cost the brigade more than 60 casualties.[45]

The 9th Brigade was to undertake its own major raid in the middle of March but 3rd Division Headquarters cancelled the operation. Monash had received instructions to prepare the Division to move to the Messines-Wytschaete Ridge section. By late April, the 3rd Division had crossed the border into Belgium, to a section of line opposite Ploegsteert Wood. It occupied the farthest right part of II Anzac Corps front, with the veteran New Zealand Division to its left.

Once the battalions had moved into their new positions, they resumed the Division's policy of aggressive patrolling and raiding. Two patrols carried out by the 39th Battalion were typical and illustrate the Diggers' determination to control no man's land. On the evening of 5 May, Private G. E. Watkins led a party into the night. Before slipping over the parapet, a sentry apprised Watkins of his suspicions that an enemy patrol was also about. Watkins moved cautiously towards the centre of no man's land, until he observed a movement in his direction. He ordered his patrol into cover and waited. A lone German approached. Watkins drew his revolver and challenged the enemy soldier, who dropped flat, then rose and surrendered. The next evening, Private C. Cameron and three other soldiers were exploring no man's land. Suddenly, a German patrol fired on them, wounding all but Private Tom Martin, who promptly launched a single- handed attack with bombs and rifle fire, driving the enemy from the position. Martin returned to his mates and carried them one by one to safety.[46]

While the 3rd Division's boldness gained it a reputation for controlling no man's land, sometimes determination went too far and resulted in high casualties. In the lead up to the Battle of Messines, Plumer arranged for his artillery to fire a number of practice barrages, to trick the Germans into responding, and reveal their defensive fire barrage plan and the location of their batteries. To take advantage of one such barrage, Monash ordered the 44th Battalion to send a raid out in its wake. Forty raiders went over in broad daylight at 2 p.m. The audacity of the operation took the Germans by surprise and the Australians gained the enemy trench, bringing back five prisoners and killing or entombing many others in their dugouts. The Germans put up stiff resistance, however, and more than half the party became casualties.[47]

The battle of Messines was a necessary prelude to the decisive battle of Passchendaele, planned by Haig for later that summer. The Messines-Wytschaete Ridge rose above the flat fields of Flanders, and from its heights, the Germans could observe and direct fire upon the British forces in the Ypres Salient, as they prepared for the offensive. To safeguard these preparations, deny the enemy observation posts for its artillery and provide a defensive flank for the coming offensive, the Second Army had to take the ridge.

Preparations had been under way for some time. During 1916, tunnellers had progressively dug a network of shafts, approach routes and chambers under the German position. Beneath its surface were 21 vaults, packed with high explosives. Four mines lay in front of the 3rd Division's zone, three beneath ground, over which it would advance, and one just to the right of the attack zone. The Second Army also concentrated twelve divisions of infantry and more than 1500 field and 750 heavy guns to support the assault while in the air 300 observation machines of the Royal Flying Corps had systematically photographed the enemy's positions.

At 11 p.m. on the night of 6 June, the assault troops began to make their way forward. Zero hour was set for 3.10 the following morning. The enemy had by chance chosen that very time to undertake an extensive gas shell bombardment on 3rd Division's positions. Although the Germans did not know the precise time at which the British intended to strike, it was impossible to hide the Second Army build-up. The bombardment which fell on the 3rd Division was only meant to interfere with the Australians' preparations. On leaving their billets, the troops were enveloped by a cloud of gas of such density, it made the approach march into a nightmare. The men put on their masks, but in the still air the gas did not disperse. Its lingering mist made the dark night even more impenetrable. In their respirators, the Diggers groped their way forward, each holding the bayonet scabbard of the man in front. Those who removed their masks to see better or to shout an order quickly became casualties. Soon gassed troops littered the approaches, and parties of lost infantry added to the confusion. Those who arrived safely found their assembly trenches filled with gas and had to remain in the open. The gas attack cost the Division more than 2000 casualties, reducing some units greatly in strength before the attack had even begun.

The 39th Battalion was one such unit and arrived at its start point with only 120 men. The 10th Field Engineers also lost heavily, commencing the battle with only 31 men out of a force of 221.[48]

Despite the gas, the remaining troops made it to their positions on time – in some cases, only just. The last company of the 34th Battalion reached its assembly point with ten minutes to spare.[49] As zero hour neared, there was a perceptible increase in tension as the Division awaited the commencement of its first major battle. Then, at 3.10 a.m., the mines detonated, ripping open the top of the ridge. Immediately the British guns roared and the infantry rose from its holes and trenches. The historian of the 3rd Pioneer Battalion described the onslaught.

> It came, and in an instant the sky was lit with a great glare in the East and North, as the mines went up. Then the earth rocked and quivered, and the roar of the explosions came, followed by the drum fire of our guns.[50]

The infantry of the 38th Battalion also felt the earth shake, and for one alarming moment, it seemed to them that 'the earth would belly up and hurl them into eternity'. Suddenly, the shaking stopped and they saw a volcano-like flame appear on the ridge, sweeping away the dim light of the dawn and flooding the countryside with blood-red light. The sound of the mine explosion reverberated through their position, only to be replaced by the scream of the guns, as the artillery opened fire.[51]

The troops quickly moved forward against an opponent stunned by the force of the explosions. Monash had decided to employ the 9th and 10th Brigades for the attack, with the 11th in reserve. Each brigade used two of its battalions to seize the enemy's forward line, while the third advanced through to the day's final objective, nearly 2000 yards away at its farthest point. The final battalion served as a local reserve.

Enemy resistance was at first almost non-existent. The easy advance proved a relief after the horror of the gas-filled approach march. The mine explosions consumed entire German garrisons and produced craters that measured 300 feet across and 50 feet deep. The leading battalions – 33rd, 34th, 38th, and 39th – had no difficulty reaching their objectives, and, in fact, some casualties were caused by men pressing forward so quickly that they advanced into their own barrage. Shortly after 5 a.m. the Division was at the crest of the ridge and digging in to repel the anticipated German counter-attack. In the afternoon the Second Army continued to advance, but the 3rd Division only had a minor role in that action.

On 31 July, General Hubert Gough's Fifth Army, further to the north in the Ypres Salient, advanced to take Passchendaele. This great battle was Haig's attempt to win the war with a single crushing blow. Haig required Fifth Army to rupture the enemy's front, push forward mobile troops, strike towards the Belgian coast and seize the enemy's rail centre at Roulers. Once the British had severed the enemy's lines of communication, the Germans would have no choice but to retreat and vacate much of the territory they had captured, or to capitulate. Haig's ambitions, however, were unrealised. The enemy's

6: Members of the 10th Machine Gun Company examine an enemy pillbox on Messines Ridge, July 1917. [*Courtesy* AWM E00555]

formidable defences prevented the British from achieving a breakthrough. Instead of the desired rapid advance, the battle declined into a series of bloody stages, as the British attempted to push their way through the enemy lines. Each attack gained for Haig a further sliver of Belgian territory, but decisive victory proved elusive.

The attack began at 3.50 a.m. and was at first successful as the attackers quickly obtained their initial objectives. However, as the British penetrated the enemy's lines, the German defences grew in strength. More significantly, the weather worsened. The rain, which had begun as a drizzle, developed into a sustained downpour lasting several days. The water soon reduced the ground to a sea of mud, making any movement extremely difficult. The wounded fell on semi-liquid ground and the mud gradually sucked them down into the suffocating earth. By the first day's end, the British were at best only half-way to their objectives, while the enemy's resistance continued to strengthen.

Also on 31 July, Second Army made a diversionary thrust towards Lille, to help Fifth Army with its attack. On the 3rd Division's front, the objective was a series of enemy outposts lying in front of a windmill, near the village of Warneton. At 3.50 a.m., the time of the offensive in Flanders, the 42nd and 43rd battalions of the 11th Brigade attacked and overran the defenders. The Australians immediately consolidated their gains, while the Division used its artillery to break up a German counter-attack. By the end of the day, they had secured the position and had dug a new line of trenches. However, this success did not come cheaply, as the German defenders had fought hard and had inflicted more than 500 casualties on their attackers. The feint was also a strategic failure. The attack had not fooled the Germans, who correctly focused their attention on the British main attack at Ypres.[52]

While the battle in Flanders destroyed divisions on both sides, the 3rd Division enjoyed a comparatively quiet time. On its sector, the Messines front, it only participated in routine operations. The relatively easy conditions improved further when, in the middle of August, the Division came out of the line and moved to a Second Army rest area near the coast by St Omer. Although the games and sports the Division's officers organised during the month long stay helped rejuvenate the men, the rest camp also had a serious purpose. Each unit underwent training in the latest developments in trench warfare, including practice in semi-open warfare, the reduction of strong-points and the importance of 'mopping-up' to eliminate any enemy soldiers who might otherwise take the assaulting wave in the rear, once it had passed. The Division also incorporated reinforcements, restoring its units to full strength.

However, everyone knew that these pleasant days by the seashore could not last. At the beginning of September, the divisional artillery broke camp and headed north. On 22 September, Haig inspected the remainder of the Division, a sure sign that the rest period was over. A few days later, Monash received orders to rejoin the war. Soon the formation was on the march to the salient, to take its turn in the killing mill of Passchendaele.

By the time the Division arrived in Ypres, the battle had been raging for over two months, and the constant pounding of the guns had destroyed virtually everything, leaving the ground a cratered moonscape. Captain W. J. Rose of the Pioneers remarked that the devastation of the country was complete. Only the slightest evidence of villages remained, and overlapping shell holes pock-marked a band of land five miles wide. The only remaining landmarks Rose could see were the enemy's concrete pill boxes.[53]

Despite the high cost, the British had enjoyed a number of limited successes which suggested to Haig the possibility of a breakthrough. Each assault had brought the army closer to Passchendaele, and, in Haig's mind, each engagement brought decisive victory closer. In September, the rains had ceased, allowing the ground to dry out, and increasing the potential for rapid movement. In early October, hoping that a few more attacks would see the collapse of the German army, Haig ordered Plumer and Gough to accelerate the pace of the offensive. This assessment of the enemy's condition was, of course, incorrect. The Germans had suffered grievously, but were not yet at the breaking point and remained capable of a determined defence.

The 3rd Division's first attack in the Passchendaele campaign took place on 4 October and became known as the Battle of Broodseinde Ridge. Due to Haig's desire to accelerate the pace of the offensive, Plumer could give Monash only three days to prepare his Division. Monash accepted that those expectations were a tribute to his Division, as only a highly competent formation could meet such a close deadline. For the offensive, the Second and Fifth Armies employed twelve divisions, attacking on a 14,000 yard front. II Anzac Corps position was in the middle of the attack zone. To the 3rd Division's left was the New Zealand Division and, to its right, the 2nd Australian Division of I Anzac Corps (with the 1st Australian Division further to the right). The four Anzac Divisions were to advance 1200–2000 yards, to secure the heights of Broodseinde Ridge. For the first time, the 3rd Division was to fight beside the other Australian divisions. Zero hour was 6 a.m.

Monash decided that the 3rd Division would attack with two brigades, the 10th on the left and the 11th on the right. The 9th Brigade was in held reserve. On the evening before the battle, the assault troops made their way forward to their assembly positions. However, in the early hours of 4 October, an enemy deserter came into the Australian lines with the information that a German attack was imminent. At 5 a.m. the Germans began bombarding the 3rd Division's front line with artillery and trench mortars. At 5.30, they increased their fire, raining shells down upon the eight battalions of infantry on the ground awaiting zero hour. As the shells exploded around them, the immediate future seemed to hold only two options for the Diggers. Either a shell would land in their particular shell hole, or zero hour would arrive and they could escape the barrage by advancing to the attack. Trapped in the open, the Division suffered many casualties. The 44th Battalion's D Company lost 50 per cent of its strength and other units also suffered heavily. Except for limited counter-battery fire, the British guns remained silent. The British planned to start their barrage at zero hour, to gain an element of surprise. In any case, at that late stage, it was not possible to reply to the German guns because it would have disorganised the artillery plan for the opening phase of the battle.

At 6 a.m. the British guns suddenly roared to life, signalling the start of the attack. Schwinghammer, of the 42nd Battalion, noted what he saw.

> ... the scene which confronted us I will never forget nor could I adequately describe. It seemed as if hell had been let loose on earth. The ground was shaking, and the air hot and full of the smell of powder from the guns. The noise was terrific. Dead and wounded were lying about and for as far as one could see on either side was a mass of soldiers moving forward, behind the barrage. Shells were falling in front and amongst us and earth was being thrown into the air. ... I saw one shell fall amongst a group of our men, seeing human limbs ... being hurled into the air.[54]

The leading battalions of the 10th and 11th Brigades – the 37th and 43rd – advanced towards the German lines, the Diggers sheltering behind a creeping barrage of shells which blasted their way through the enemy's line. To their surprise, as they moved forward, the Australians met the enemy in strength in no man's land. The deserter's warning had proved correct, and the German barrage had actually been a preliminary to the enemy's own attack. The Australians had beaten them to the punch by only ten minutes, and the advantage rested with them as their creeping barrage rolled through the massed German ranks posed at the point of their own offensive. Lewis gunners began to fire, and riflemen lunged with bayonets as the two opposing lines collided in fierce hand-to-hand struggle. The ferocity of the British barrage, combined with the unexpected appearance of the

Diggers, crushed the enemy's resistance. Soon German troops began to surrender in mass while others fled to the rear only to be overwhelmed once again by the British barrage.

With the advancing infantry were a number of specialists from the supporting arms. Mortar men from the trench mortar batteries brought their weapons to bear against enemy strong points, while Lewis gunners from the 3rd Pioneer Battalion added their firepower to the assault. One Pioneer, Lance Corporal W. Peller, received the Victoria Cross for his actions. He was among the first wave, when he encountered an enemy party sniping at his mates. He rushed their position, Lewis gun blazing, and accounted for nine Germans. He repeated this act twice more, on each occasion clearing the way for the advance. He then turned his weapon on an enemy machine gun nest and killed the crew. His total haul for the day was 30 Germans.

Within half an hour, the leading battalions had reached their objectives and the following units – the 38th and 42nd Battalions – leapfrogged through to the fore, to continue the attack. At each transition, the creeping barrage would pause to allow the next battalion through, and then, at the prescribed time, the shells marched forward again and the infantry continued to advance behind their shelter. The 39th and 44th Battalions followed in turn, clearing out enemy strong points as they went. The 40th and 41st Battalions tackled the last phase. Although German resistance had begun to stiffen, the Australians were still able to drive the enemy from their strong points or kill them inside. In one instance, two Tasmanian sergeants from the 40th Battalion – H. Boden and S. J. Barrett – approached a particularly large pillbox. Boden fired a shot through its loophole, an action which brought forth a stick with a piece of white cloth attached. The two accepted the surrender of 70 Germans who had had enough. By 9.15 a.m., the Division began to dig a new line along the top of the ridge, 2000 yards from their starting point.[55]

In the afternoon, the Germans attempted to organise a counter-attack on the 11th Brigade's and 2nd Australian Division's fronts, but the Australians broke it up before it could develop into a serious threat. From the ridge's summit, artillery observers had a clear line of sight and accurately directed fire onto the Germans below. Providing closer support, Lieutenants C. W. Couchman and J. W. Howie of the 11th Trench Mortar Battery prowled the 41st Battalion's front line, firing their bombs onto parties of Germans wherever they presented an adequate target.[56]

Although the gains had not come cheaply – the Division had suffered more than 1800 casualties – it was clear that the attackers had achieved a major success. Enemy losses were far higher, and the Australians had rounded up large numbers of prisoners. The 40th Battalion alone had sent back 300.[57] Furthermore, the Division had wrested from the enemy a well designed and strongly fortified position. In a letter home, Spencer E. Jenkins, of 3rd Divisional Signal Company, described the strength of the enemy's defences. He observed that the German pillboxes were enormous structures with concrete roofs at least five feet thick and reinforced with iron bars. The enemy, Spencer continued, had built these as part of a defensive system consisting of a number of lines. Each strong point lay within supporting range of another, about 300 yards distant, while approximately three-quarters of a mile separated each line. Spencer also wrote that German bodies choked the ground around his position, fresh casualties from the recent attack and older ones the artillery had blasted back to the surface. Marvelling at the Division's accomplishment, Spencer remained supremely confident that the Australians would continue to best the enemy.[58] Monash shared his men's belief that they had accomplished a remarkable feat. He cabled home the news: 'All well [,] division again brilliantly victorious in the "greatest battle of the war"'.[59]

Three days later, Godley rotated the 3rd Division out of the line for a brief rest before its next operation. The Second Army attacked again on 9 October, but this time Godley used II Anzac Corps' British formations – the 49th and 66th Divisions. However, before the British could launch the next blow, the weather changed again. On 5 October, the rain returned. It began as a gentle mist, grew into showers and by 8 October had escalated into a torrential downpour.[60] Drainage on the shattered battlefield was an impossibility, and the ground quickly reverted to a quagmire of mud and water-filled shell holes. These conditions contributed to the German repulse of the British on 9 October, when the attackers failed to gain ground and suffered heavy casualties. When the 3rd Division returned to the line on 10 October, in anticipation of an attack on the 12th, the battlefield had not improved.

For the 3rd Division's next battle the objective was the Passchendaele ridge itself. II Anzac Corps was again in the middle of the attacking line, with the New Zealand Division on the 3rd Division's left, and the 4th and 5th Australian Divisions of I Anzac Corps on its right. British formations of Fifth Army extended the line further to the north, while the British X Corps supported the southern flank. This time, Monash decided to attack with the 9th and 10th Brigades, holding the 11th in reserve. Each brigade led with one battalion, while the other followed through to seize successive objectives. The division's destination 3000 yards ahead was the key to the defender's position – the top of Passchendaele ridge.

The conditions for the renewed attack, however, were not auspicious. The poor state of the ground impeded preparations and, critically, had slowed down, and in some places prevented, the forward movement of the artillery and ammunition dumps after the gains achieved by the Broodseinde battle. As the 3rd Division readied to attack on 12 October, many of its guns were out of position and short of shells. The War Diary of the 29th Battery explained that the mud was so deep that the trip from the wagon lines to the battery position, only an hour on stable ground, was a seventeen-hour slog. Animals floundered in deep shell holes and men often sank so deeply that, when pulled out, their boots and breeches remained behind. As a result, when the 3rd Division attacked on 12 October, it had significantly less fire support than the artillery had provided for the Broondseinde battle.[61]

The assaulting units began to move into their jump-off positions after dark on 11 October. It was a difficult passage. Monash had assigned 10th Brigade 'K' track, and four battalions had to make their way forward, single file, over its surface. The track was in relatively good condition with duckboards covering most of its length. However, it was a twisting and narrow path, and the mud of the adjacent terrain was between one and four feet deep. As well, the track lay in clear view of the enemy during the day so that, at night, they knew its position and rained shells down on the column. Periodically, one would land amid the column, causing casualties and shattering the duckboards. The following troops then had to wade into the mud, across the crater, to where the path resumed. Lieutenant G. M. Carson of the 33rd Battalion wrote that his men had to march through a sheet of iron. Even before the attack commenced, he noted, his unit had lost scores of men. The German gunners had also sent over gas shells, forcing the troops to don their respirators which added to their difficulties. Although most units were in position by around 3.30 a.m., Carson's battalion reached its jump-off point with only four minutes to spare before the battle's commencement. As the last unit in the 9th Brigade's column, it had spent more than seven hours on the approach march.[62]

Although not part of the assaulting wave, the stretcher bearers of the 9th Field Ambulance moved up behind the infantry to assist the wounded. The unit historian provided a description of their approach march:

> After the main road had been left behind the way lay over an area that was merely a succession of shell holes, and the guiding white tapes twisted and turned like some long snake, as their placers had endeavoured to pick a way where practically no way existed. And all the time, without ceasing, or easing for a minute, down came that hail of enemy shells.[63]

Although the deep mud made the trip an ordeal, it also allowed the troops to survive a little longer, as the slurry-like ground swallowed many of the enemy's shells and prevented their detonation. Reaching their destination, the stretcher bearers found pillboxes, and even shell holes, filled with wounded left from the 66th Division's attack of a few days earlier. Instead of enjoying a few hours rest, the bearers set to work retracing their steps to the aid stations.[64]

At 5.25 a.m. the British guns began to fire, signalling the attack. It was immediately apparent to the infantry that something was wrong; the creeping barrage which rolled out in front of them was much thinner than it had been for the Broodseinde attack. Private J. Hardie, who went over the top with the 9th Machine Gun Company, observed that the advance had only just begun when German shells began to explode in the Division's ranks. Battalion after battalion were practically cut to pieces. Hardie could not explain how his team got forward but, after 500 yards, enemy fire stopped their advance. Wounded, Hardie made his way to the rear and on to Blighty to recover.[65]

The mud proved the critical factor in the battle. As Carson attempted to advance, his men sometimes became bogged up to their armpits. Others, he recalled, were drowned in the mud and water. Because of the mire the infantry was unable to maintain the required rate of advance of 100 yards in four minutes and, as a result, even the limited creeping barrage was soon far ahead of them. Unmolested, the Germans could now direct their weapons without threat at the Australians from strong points and trenches. By 7 a.m. the

7: The mud of Passchendaele. Mules bring supplies forward for the 42nd Battalion, October 1917. [*Courtesy* AWM E00952]

British guns were practically silent, while the enemy's artillery continued to sweep the attack zone. The gunners had not been able to move forward enough guns, and had insufficient ammunition for those pieces they could use. Nor had they had time to build proper gun platforms and each time an artillery piece fired, the recoil pushed it into the mud, burying it hopelessly. As the day progressed, the infantry was increasingly on its own.[66]

The 10th Brigade managed to get as far as its first objective and the 9th Brigade took its second objective, but the troops could get no further. More troubling, as the morning progressed, the Australian situation grew increasingly untenable. The New Zealanders, on the left, had been stopped soon after leaving their start line by an impassable swamp and the murderous fire that poured down on them from Bellevue Spur. This caused a gap to open between the two divisions, and enabled the Germans to enfilade the 10th Brigade's position with deadly fire, while enemy reserves moved up to counter-attack the Australians on their exposed flank. On the right flank, another gap had opened up between the 3rd and 4th Australian Divisions. Furthermore, the enemy's artillery, virtually unopposed by the British, could fire unmolested. From the heights of the Passchendaele Ridge, German artillery observers accurately plotted the progress of the Australians and directed devastating fire onto their positions. The 3rd Division had no choice but to retire and form a new line just in front of its starting positions. That evening, the 11th Brigade came forward to relieve those who had participated in the attack. The battle cost the Division nearly 3200 men.[67]

12 October marked the end of the Division's part in the Passchendaele battle. On 22 October, it left the salient. The Canadians took over the role as spearhead for the rest of the campaign. After a pause to allow the ground to dry out and to prepare their guns properly, the Canadian Corps resumed the offensive in early November. It took them two more attacks, with full artillery support, to achieve the objectives set for the Australians on 12 October. After reaching the top of Passchendaele Ridge on 11 November, Haig brought the offensive to an end. After more than three months of battle, victory eluded him. The British would have to make another effort the following year.

The 3rd Division at War: 1918

The 3rd Division spent the winter of 1917-18 guarding quiet sectors of the front or resting behind the lines. Between November and March, the formation found itself on familiar ground, reacquainting itself with the defences of Armentières and Ploegsteert or serving as the corps's reserve division. During this period, the administration of the Division changed. Haig acceded to a long-standing request of the Australian Government that its divisions be united into a single national force. The result was the replacement of the two Anzac Corps with a single Australian Corps. As the Australian divisions withdrew from Passchendaele, they entered the new command arrangement. The 4th Division spent the winter serving as a depot for the other Australian formations but, in the spring, joined the rest of the Corps. Not surprisingly, the command of the Australian Corps went to General William Birdwood, who was well liked by the men. Birdwood had had an association with Australian troops since 1915, when he succeeded General Bridges as General Officer Commanding-AIF, and had also commanded I Anzac Corps since its raising.

There was no doubt that the British would attack in 1918. Although the Passchendaele campaign had been a nightmare for those who fought it, its effect on the enemy had been no less severe. Haig believed that a few more blows would see the end of the enemy's resistance. However, during the winter, it became increasingly obvious that the British and French would first have to withstand a German offensive. The enemy was in a desperate strategic situation. In early 1917, the German navy had resumed unrestricted submarine warfare against merchant shipping, in an attempt to starve Britain out of the war. The U-boats failed to achieve that goal but, as a direct result of their sinking American merchant ships, the United States entered the conflict on the side of Britain and France. Although unprepared for war in 1917, American troops began to arrive in France in overwhelming numbers the following year, and their presence would decisively shift the manpower ratio in favour of the western allies. General Erich Ludendorff realised that Germany had to end the war before the United States could bring its strength to bear. The collapse of Russia in late 1917 allowed him to transfer some troops from the east to France, giving him a temporary manpower superiority on the Western Front.

Ludendorff made his bid for victory on 21 March 1918, near St Quentin in Picardy, along a broad section of front held by Gough's Fifth Army. Gough's over-extended army collapsed and German divisions advanced into the Somme valley, tearing a huge hole in the Franco-British line. Enemy troops succeeded in penetrating the Fifth Army's defences, pouring forward into the open fields in a display of mobility not seen since the opening campaign of the war. As the Germans advanced into the British rear, they drove a broad wedge between the two allies and threatened the very survival of the Franco-British alliance.

When the Germans attacked, the 3rd Division was already in motion, not to the Somme, but towards Ypres, where it was to move into reserve. It had been in a rest position along the coast when General Headquarters had ordered it to Flanders to support Second Army, in anticipation of an attack against that strategic sector. Ludendorff's decision to attack near the Somme took the British command by surprise. Monash himself was away at the time, enjoying a well earned rest in the south of France.

On 24 March, as the scale of the German breakthrough became apparent, Haig ordered the 3rd Division south to the Somme. In particular it was to move into the German path and stop the enemy from capturing Amiens. This city was a vital rail hub and supply centre, and its loss would force the British to fall back towards the channel ports, separating them from its ally France, which would have to retreat towards Paris. Within a few hours of receiving the new orders, the 3rd Division had changed direction and was moving to stem the tide. Monash, realising the significance of the threat, was on his way back to rejoin the Division.

The unity of command established by the creation of the Australian Corps was now temporarily ignored, and Haig's staff placed Birdwood's divisions under the control of other corps. Haig transferred the 3rd Division to the VII Corps under the command of Lieutenant-General W. N. Congreve. The Division was to concentrate immediately to the east of Amiens, directly in the path of the German advance. Monash had now returned and, on the evening of 26 March, arrived at the VII Corps Headquarters in Montigny, a little village to which Congreve had recently and hastily retreated. Congreve's interview with Monash was brief and to the point. He explained that the enemy had attacked again in the afternoon and that his three weak and exhausted divisions had given way. He ordered Monash to move the 3rd Division into a tongue of land between the confluence of the Ancre and Somme Rivers, to get as far east as possible and to stop the Germans. He also issued equally desperate orders to Major-General Ewen Sinclair-MacLagan, the General Officer Commanding 4th Australian Division, instructing him to place his troops on the heights above the Ancre, to the left of Monash.

At daybreak the next morning, accompanied by a single staff officer, Monash reached the town of Franvillers, where he intended to establish his headquarters. He was the first member of the Division to arrive. Through his glasses, he could see German cavalry approaching from Morlancourt, only 3 kms away. The only friendly force between him and the enemy was a composite unit under the command of Brigadier H. R. Cumming, containing the few remaining British infantrymen and cavalrymen capable of resistance. Congreve had instructed Cumming to delay the enemy and to cover the arrival of the Australians. Later that morning, worn-out motor buses pulled into Franvillers carrying two battalions of the 11th Brigade. Monash commented on the sight:

MAP 3. The 3rd Division before Amiens

> The spectacle of that Infantry will be ever memorable to me, as one of the most inspiring sights of the whole war. Here was the Third Division – the 'new chum' Division, which in spite of its great successes in Belgium and Flanders, had never been able to boast, like its sister Divisions, that it had been 'down on the Somme' – come into its own at last, and called upon to prove its mettle.[68]

Without a moment's delay, Monash ordered his troops to cross the Ancre and establish a line running from Mericourt on its bank to Sailly-le-Sec on the Somme. The German advance on that front was about to come to an end.[69]

As the men of the 3rd Division arrived in the Amiens sector and moved forward to their positions, the scale of the calamity which had befallen Fifth Army became apparent. Fleeing refuges choked the roads and empty villages showed evidence of hasty departures. The Australians passed broken bodies of British troops making their way to the rear, who warned the Diggers that they would soon be doing likewise. Instead of weakening the

Australians' resolve, such comments only reinforced their determination to get on with the job.[70] Schwinghammer forced his way forward with the 42nd Battalion, observing that the road was blocked with old men, women and children, pushing carts or wheelbarrows filled with their possessions and driving livestock before them. As his battalion descended into the Ancre Valley, it passed through the village of Heilly, which lay abandoned. Its inhabitants had fled with such haste that their tables were still set for a meal which the hungry troops consumed as they marched through the village. Passing a priest, Schwinghammer noted that he blessed the Australians as they went by. The 42nd pressed on to Sally-le Sec, where it dug in, while the 43rd Battalion occupied positions to the left.[71]

It was clear to Monash that the 3rd Division stood alone. Behind him, there were no reserves to stop the Germans if his men failed. He sent the 44th Battalion into Corbie during the night of 27-28 March, with orders that the unit was to defend to the last man the town's bridge across the Somme.[72] Worsening the situation was the Division's lack of firepower, and although further units of infantry arrived in Franvillers on 27 March, the artillery did not reach the area for another two days. The only guns at the Division's disposal were a few 13 pounders belonging to a British cavalry division.[73] As a precaution against disaster, the 10th Field Company prepared the bridges across the Ancre River for demolition, while the 11th Field Company did the same for those across the Somme. Later, on 6 April, as the Germans approached, Captain G. L. A. Thirkell of the 11th Field Company dropped the Bouzencourt Bridge into the Somme Canal.[74]

Despite its weakness, the 3rd Division lost no time in letting the enemy know of its arrival. At 8 a.m. on 28 March, the 38th Battalion took over from the 43rd and immediately pushed forward to the far side of Mericourt. Captain F. E. Fairweather then advanced his C Company a further 1000 yards and occupied Marrett Wood. Shortly thereafter, a party of about 100 Germans crossed the crest of a hill in front of C Company's position to challenge the seizure of the wood. Fairweather's men opened with intense rifle and Lewis gun fire, cut down half the enemy and forced the rest to flee. That night, Monash ordered his entire line to advance. Near the Somme, stiff resistance from the village of Sailly-Laurette held the Australians to negligible gains, but on the left, along the Ancre, the Division occupied the village of Treux.

On 30 March, the Germans made a serious attempt to force their way through the Australians and resume the march to Amiens. On the previous day, observers in the 44th Battalion had noted the arrival behind the enemy's position of buses full of troops and columns of marching infantry, and spent the night preparing for a German attack at first light. To the surprise of all, when dawn broke, the German lines remained quiet. The Australians settled down and spent the day resting and watching the enemy. Suddenly, at noon, the enemy opened a terrific bombardment on the Australian position, although most of the shells fell wide. To the rear of the Australian line, the fire soon reduced to rubble the village of Sailly-le-Sec, which so far had not known war. As the guns blazed, masses of German troops poured out of Sailly-Laurette and advanced, in broad daylight, towards the Australian defenders – the 42nd and 44th Battalions of 11th Brigade. Alerted by sentries, dozing Diggers grabbed their weapons. As the enemy descended the hill in front of Schwinghammer's position, the Germans presented a perfect target. Australian rifle and machine gun fire poured into the German ranks, and soon the ground was thick with enemy dead and wounded. Within a few minutes, the Australians had broken the attack. During the afternoon, the Germans made a series of further assaults, each with the same result. By the day's end, Schwinghammer estimated that thousands of German bodies choked the hillside. After the war Monash chose Sailly-le-Sec for the location of the 3rd Division monument which commemorates its tenacious defence during those days in front of Amiens.[75]

The Germans made a number of further efforts towards Amiens. 10th Brigade repulsed an enemy attack at Dernancourt on 6 April – but the Australians did not budge. Ludendorff's attempt to break through and capture Amiens had failed. During the spring, the Germans launched a series of further offensives on other sections of the Western Front; however, none proved as serious as the initial blow which the 3rd Division had helped stop.

Having halted the enemy, the Australians now seized the initiative and switched to the offensive. During May, in a series of small battles, they pushed the enemy back a few miles, widening the safety zone in front of Amiens. At this time, Monash instituted a policy he termed 'peaceful penetration': the nightly dispatch of small bodies of men, even individual soldiers, into no man's land to inflict casualties, take prisoners and demoralise the enemy.

On 21 June 1918, the 33rd Battalion undertook such an operation. Five men, under the command of Lieutenant A. M. Widdy, left their trench at 5.45 p.m., just prior to the commencement of a bombardment on a position to the immediate right of their objective. They made their way through a belt of high wheat which obstructed no man's land at that point, while the noise of the guns provided a useful cover. The wheat ended about fifteen feet from the German parapet, and the raiders used its cover to observe the enemy's position. Widdy spotted a fire bay where two Germans were sheltering and decided to try to capture them. He covered the short distance and leapt into the trench, followed by Corporal A. W. Brecht. The Germans resisted and Widdy killed both, despite the fact that one had got hold of Brecht's rifle. Three more Germans appeared from a nearby dugout; Brecht bayoneted one. The other approached Widdy, revolver drawn, and fired, hitting Widdy. Despite the wound, Widdy soon despatched his foe. By now, the fifth German tried to flee, but Brecht shot him. The Australians then heard an enemy machine gun team approaching and decided that it was time to leave. They made their way out to the wheat and safely back to their lines, while the enemy's machine gunners tried in vain to find them. As well as four dead and one wounded, Widdy brought back important intelligence. He reported that the enemy wire was in such poor condition that it did not offer a real obstacle, and that their trenches were so narrow that it was impossible for the rest of his patrol, who had also jumped in, to take part in the melee.[76]

During this period, command changes within the British structure affected the command of the 3rd Division. After the debacle which befell Fifth Army, Haig sent

8: The 10th Infantry Brigade in front of the remains of Treux Woods after its capture, 28 March 1918. [*Courtesy* AWM E01955]

Gough home and replaced him with Birdwood. At the end of May, Birdwood's departure in turn led to the promotion of Monash to the rank of Lieutenant-General and to the command of the Australian Corps. His replacement as General Officer Commanding 3rd Division was Major-General John Gellibrand.

Now in charge of the Australian Corps, Monash had five divisions under his command. That corps, as a part of General Henry Rawlinson's Fourth Army, was to take a leading role in the British Army's march to victory in the war's closing months. Monash sensed that the Germans were weakening, and that the opportunity existed to resume the offensive on a scale larger than trench raids. He proposed to Rawlinson that his forces seize Hamel, a small village south of the Somme River on the Amiens sector, and Rawlinson approved the operation. At 3.10 a.m. on 4 July, elements of the 2nd, 3rd (11th Brigade) and 4th Australian Divisions, supported by 60 tanks and some American infantry, took the enemy by surprise and seized the town. Although resulting in a gain of only two miles, the Australian success at Hamel, combined with a major French victory at Soissons on 15 July, reinforced the recognition that the German Army was vulnerable and demoralised.

Haig believed that the opportunity for decisive victory had arrived, and authorised Rawlinson to plan a major attack. Fourth Army would soon strike again and achieve a

9: Forward observers from the 11th Machine Gun Company watch enemy movements near Sailly-le-Sec, 29 March 1918. [*Courtesy* AWM E01846]

tremendous British victory at the Battle of Amiens, which Ludendorff rightly described as Germany's blackest day. That victory opened the war's final campaign, and, in a series of highly successful attacks, the Australian Corps (including the 3rd Division) were at the forefront of the British army's efforts.

The Battle of Amiens began on 8 August 1918. Rawlinson had at his command three corps. On the left was the III Corps, composed of British troops, the Australian Corps held the middle and the Canadian Corps was on the right. The French First Army, to Rawlinson's right, would also participate in the attack. Monash planned to employ the 2nd and 3rd Divisions to seize the Australian Corps's first objectives. When they had reached their goal, the 4th and 5th Australian Divisions would pass through and continue the offensive to its final objectives. The 3rd Division was responsible for breaking through the crust of the enemy's defences, so that the follow-on waves could strike deeply into the enemy lines. Gellibrand would use the 9th and 11th Brigades in the assault, with the 10th in reserve.

At 4.20 a.m. on 8 August, the guns of Fourth Army, previously silent except for routine firing, suddenly unleashed a massive bombardment on the enemy's positions. Ken Murray, a private in the 38th Battalion, wrote in a letter to his father that the troops had awaited

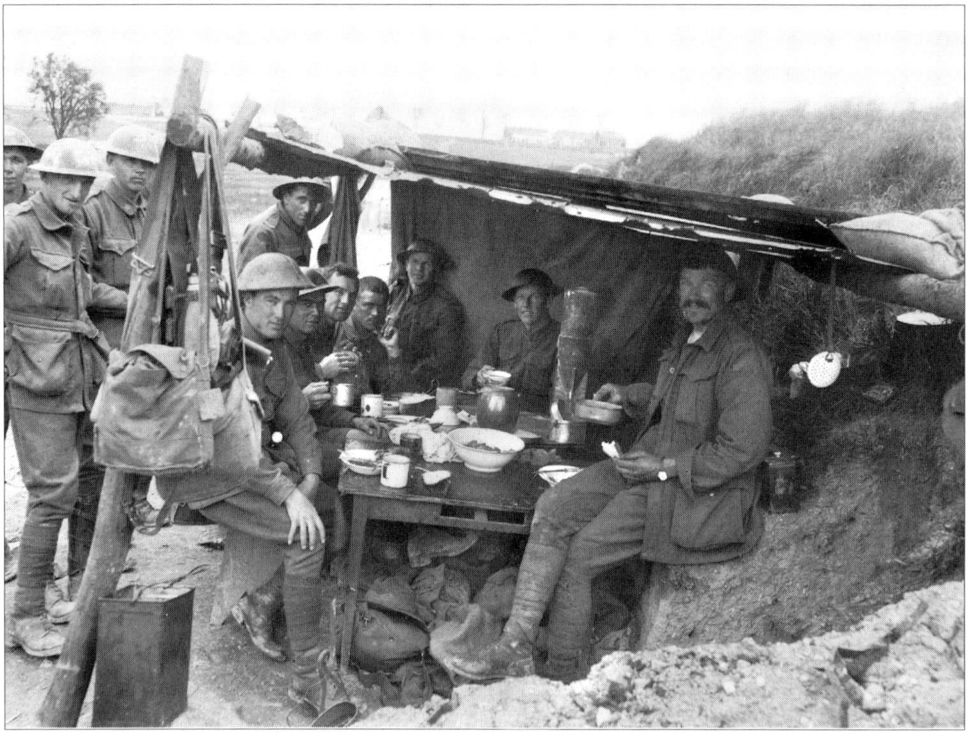

10: Members of the 6th Medium Trench Mortar Battery enjoy a meal near Villers-Bretonneux, 20 June 1918. [*Courtesy* AWM E02557]

the bombardment with 'breathless anticipation':

> It was dark and foggy – and very very still. At last, out of the fog, came a deep 'rumble-rumble' [then] a sharp 'bang-bang', then with the most terrific crash every gun on a front of thirty miles fired simultaneously.

He continued that it was useless to speak and impossible to be heard, and that the ground shook beneath his feet and his brain was hammered by a series of sound waves. To Corporal J. S. Finney in the 44th Battalion, the opening barrage was a magnificent sight. As far as he could see, to the right and left, red flame filled the sky. Intermingled were multi-coloured bursts of the enemy's Verey lights, as German units called in vain for assistance from their own guns. The British struck the enemy's batteries with such an avalanche of shells that the enemy gunners were unable to stem the advance. The power of the attack was even more evident to Murray the next day when he examined Accroche Wood, one of the Division's objectives. Before the battle, Murray observed, it had been a green and shady wood. Now it was stripped of leaves and the trees lay shattered like matchwood.[77]

With the sound of the guns, Australian, Canadian and British soldiers rose and set out across no man's land, behind a creeping barrage of bursting shells and in the wake of 400 tanks which Rawlinson had used to crush the enemy's barbed wire and roll over any enemy machine guns which dared to fire. As the troops advanced, the artillery fired gas and high explosive shells onto the enemy's batteries, swiftly silencing almost every German gun. The Diggers moved quickly into the enemy lines. As Finney's platoon moved forward, it encountered a party of enemy soldiers, their hands up crying 'Kamerad! Kamerad! Mercy!' The Diggers approached them carefully, suspicious of a trick. It was soon evident that the Germans had no fight left in them, and the Australians directed them to the rear.[78]

By the day's end, the attackers had reached all their objectives, except on some parts of the III Corps front. The historian of the 3rd Division Pioneer Battalion correctly observed that for them the battle was 'the quietest day that the unit had experienced for some time, and it was hard to realise that it was the biggest operation in which they had been engaged'.[79] By nightfall, the Fourth Army had captured more than 17,000 Germans and more than 400 guns.

After Amiens, it was clear that the Australians had not simply shattered Accroche Wood, but had dealt the enemy a tremendous defeat. Australian troops now had no doubt as to who were the superior soldiers and lost few opportunities to demonstrate that mastery as they continued to engage the enemy in patrols and raids. Even daylight did not protect German soldiers from Australian aggressiveness. A typical adventure occurred on 25 August, when a daylight patrol of the 44th Battalion entered the German lines near Bray. Under the command of Captain M. Lewis, the patrol persuaded 35 Germans to surrender and marched them into captivity. The 38th Battalion's Ken Murray related another incident which provided further proof of the superiority of the Australian soldier over his formerly stalwart foe. His platoon had received orders to seize an enemy machine gun post 600 yards in front of its position. In a series of quick rushes, men covered the distance without a single casualty. When within striking range, Murray's platoon charged the post, killing four Germans and capturing one officer, 20 men and three machine guns.[80]

Haig also realised that the Germans were a beaten force, and, to prevent their recovery, ordered his armies to attack. With their French, American and Belgian allies, British forces rained unceasing blows on the enemy, forcing the Germans to retreat and demoralising them even further. The pattern for the rest of the war was semi-open warfare. Whenever the Germans attempted to make a stand, the attacking troops drove them from their defences, and the enemy then retreated to a new position where they attempted to organise a defence. On each occasion, the attacking troops drove the enemy from its defences and back towards Germany. In a series of major and minor battles, the 3rd Division and the other Australian formations were at the forefront of this final campaign.

On 21 August, Third Army, to the north of Fourth Army, began the Battle of Albert. On 26 August, the First Army attacked. The 3rd Division resumed its advance on 22 August, fighting its way to the outskirts of Bray. Once the 1st Australian Division had secured the Chuignes Valley, immediately south of Bray, the 3rd Division advanced again, sweeping through the village. Monash reported that, with only trifling loss, the 40th Battalion brought in 200 prisoners. On 25 August, the 3rd Division again pushed forward, against an obviously worn-out foe. For the next week, the Division continued to push the enemy back, taking Clery on 30 August and Allaines on 2 September. Elsewhere on the Australian Corps front the 2nd Australian Division captured Mont St Quentin, and the 5th Australian Division seized Peronne. These actions, combined with attacks elsewhere, breached the defensive line upon which the Germans had intended to make a stand and made their further retreat inevitable.[81]

Ludendorff now decided that his forces must retreat to the prepared defences of the Hindenburg Line. The Australians followed, keeping up the pressure and attacking the German rearguard at every opportunity. The action at Tincourt on 6 September was typical. In the vanguard of the Division was the 11th Brigade, with the 42nd Battalion and 3rd Pioneer Battalion leading. Gellibrand was using the Pioneers as infantry, as almost continuous employment since the end of March had greatly reduced the strength of the battalions. The Pioneers moved forward just after midnight on the morning of 6 September. Advancing slowly but steadily across a defensive system of old trenches and belts of wire, they first made contact with the enemy at the village of Buire. The Germans quickly withdrew and the Pioneers probed forward, until they again contacted their opponent at Tincourt. There the Germans occupied a strong position, and they were determined to make a stand. The Pioneers pressed forward by short rushes and occupied a ridge overlooking the village. Meanwhile the 42nd Battalion had captured a copse on the right, outflanking the enemy's defences. The defenders set fire to Tincourt and again retreated while the Australians passed through the village and resumed the pursuit.[82]

After the fall of Tincourt, Schwinghammer described a party of Germans which surrendered to the 42nd Battalion:

> They were starving and ate ravenously of the food we gave them. Their clothes were in rags and their boots worn out ... and they were also very dejected and downhearted. They knew that they were losing the war. What a contrast to our own troops. We were well clothed, well fed and full of optimism.[83]

While there was no longer any doubt as to who were the superior soldiers, Australian units were also worn out, after six months of nearly continuous fighting. The infantry, in particular, was greatly under strength, owing to the lack of reinforcements. In September, Monash attempted to break up some of the 3rd Division battalions and reassign their men to other infantry units, to achieve effective numbers. The battalions selected, having fought together for nearly two years, strenuously resisted his efforts. The men were determined to see the war to its conclusion in the units in which they had begun. A committee in the 42nd Battalion studied the King's Regulations and military law seeking a means by which they could properly disobey the order to disband. On 21 September, their commanding officer paraded the unit and announced the decision to disband the unit and transfer the men into the 41st Battalion. Upon receiving the order to march off, the officers responded but the men did not budge. A second attempt on 25 September ended the same way. Monash overlooked the insubordination and conceded the point. He postponed the reorganisation and agreed that the affected battalions could participate

11: 4.5 inch howitzers of the 108th Battery firing in support during the 3rd Division's advance, 26 August 1918. [*Courtesy* AWM E03114]

intact in the coming offensive. Disbandments would occur, but not until after the Division had fought its final battle.[84]

By the middle of September, Fourth Army had reached the outskirts of the Hindenburg Line. This defence was the most formidable fortification system created by the Germans on the Western Front. It was begun during the winter of 1916–17 and had been improved ever since. It consisted of a series of trench lines and strong points supported by belts of barbed wire, which in some places reached back hundreds of yards. Further strengthening the position was the presence of the St Quentin Canal, immediately in front of the main line, its steep banks creating an impassable barrier to tanks and a considerable obstacle to infantry assault. The only break was a two-mile gap near Bellicourt, where the canal ran through a tunnel. However, the Germans had incorporated the tunnel, too, into their defence plans; it served as a bomb-proof shelter and could harbour thousands of troops. After overrunning the British defences during the March 1918 Offensive, the Germans improved the Hindenburg Line by incorporating into the system the British trenches which lay immediately to its west. Although those trenches were not as strong as the main defence line, German engineers strengthened them by fortifying villages and making other improvements. At the time the 3rd Division approached the position, the German defences consisted of a defensive zone 6000 yards deep, incorporating six lines of trenches as well as the canal.[85]

On 18 September, Fourth Army commenced the initial phase of the attempt to breach the Hindenburg Line. On the Australian Corps front, the 1st and 2nd Australian Divisions captured the enemy's outer system of three trench lines – the former British trenches – opening the way for the assault on the main defences, which Rawlinson had scheduled for 29 September. However, the III Corps which attacked on the Australians' left failed to penetrate as deeply, and much of the German line in that sector remained in enemy hands.

Rawlinson assigned the central role in the attack to the Australian Corps and gave Monash the responsibility for planning the offensive. To help the Australians, Rawlinson also assigned to Monash two American Divisions, the 27th and 30th. Monash planned to attack straight through the heart of the German defences, namely the section of the Hindenburg Line where the St Quentin Canal ran underground through the tunnel. For that reason, the Australian Corps was shifted slightly to the left, displacing III Corps, so that its front directly faced the area of assault. When the Australians and Americans attacked over the Canal, IX Corps to its right would use the 46th Division to force a crossing, with the 32nd Division following, while III Corps on Monash's left would assist by establishing a defensive flank.

III Corps's failure on 18 September to seize the enemy's outer defences now became an annoying complication for Monash, as he finalised the plan for the coming breakthrough battle. To relieve the situation, the Americans made another attempt on 27 September to gain the required territory. Although initially they were successful, German counter-attacks later that day regained a number of key positions and left the precise demarcation of the front line unclear. The situation on the left side of the attack zone thus created a difficult problem for Monash. The right half of the Australian Corps attack was at its correct start line while, on the left, the attackers remained about 1000 yards behind the intended jump-off position. The artillery preparations were already well advanced, including the positioning of batteries and ammunition dumps and the design of the creeping barrage, and Monash and Rawlinson believed that it was impossible to make any drastic changes without postponing the date of the offensive. Monash therefore decided that the attack would proceed as planned, even though it meant that on the left the assaulting troops would have to advance about 1000 yards before they reached the assigned start line and the protection of the creeping barrage. This decision was a compromise; it allowed the attack to go ahead, and thus denied the enemy time to strengthen its defences and bring up reinforcements. But it meant that the assaulting troops would be terribly exposed to the enemy's defensive firepower until they could reach the safety of the creeping barrage.[86]

Monash planned his attack with the American troops in front. Their objective was to cross the canal, breaching the main Hindenburg Line and Hindenburg Support Line, and penetrate the enemy's position to a depth of 4000 yards. The 3rd and 5th Australian Divisions would then pass through the Americans and assault the system's final position, the Beaurevoir Line. The British 46th and 32nd Divisions would widen the breach to the south of the tunnel. Once having broken the defences, the attackers would expand the hole, rendering the Hindenburg Line untenable and forcing the Germans to retreat. Preceding the assault, Fourth Army planned a four-day preliminary bombardment in order to blast holes through the German entanglements, eliminate their strong points, and neutralise their artillery. The British would also employ mustard gas, firing more than 29,000 shells, and Monash would have at his disposal nearly 200 tanks.

At 6.20 a.m. on 29 September, the Americans 27th Division attacked in a thick mist, and immediately ran into trouble. As the Americans advanced, machine gun fire from the unsuppressed strong points in the German outer defences swept through their ranks, cutting units to pieces and forcing the survivors to seek shelter. As they struggled forward, they failed to mop up enemy dugouts in the area through which they passed, and thus soon found themselves under fire from both front and rear. Thus, when the 3rd Australian Division followed at 9 a.m., instead of passing easily through, they also came under intense German fire. Lieutenant Phillips reported to 10th Brigade Headquarters that, since the Americans had not mopped up, the Australians were receiving 'M. G. fire galore'.[87] The 3rd Division now also had to clear the enemy from the zone that had been the Americans' responsibility, slowing their own advance.

The advance to the right had gone somewhat better. The 30th Division and 5th Australian Division had secured the southern end of the tunnel and captured the main and support trenches of the Hindenburg Line, but failed to reach their ultimate objective, the Beaurevoir Line and the open fields beyond.

Monash's plan lay in ruins. When battle resumed the next day, he incorporated changes into his plan which demonstrated his great ability as a corps commander, as well as the competence of his soldiers. The Australians shifted the axis of attack so that the main thrust was now north along the main Hindenburg and Hindenburg Support Lines, from positions the 5th Australian Division had captured, instead of eastwards across them. Monash inserted the 3rd Division's reserve formation, the 9th Brigade, into that sector, to strike, with elements of the 5th Australian Division, northwards up the German trench line, while the rest of Gellibrand's division continued to apply pressure along the eastern face of the enemy's defences. Far from being a methodical advance in a single day, the battle degenerated into countless minor engagements, as small parties of Australians and Americans closed upon the enemy with bombs and bayonets and fought their way forward against a foe desperately contesting every yard of ground.

Gellibrand described the next few days as the most difficult of his experience.[88] However, the enemy was unable to check the irresistible advance of the 3rd Division and the Diggers left dead Germans in every strong point and trench bay they captured. By 1 October, it was clear that the defenders were weakening, and the 37th Battalion reported that the enemy was evacuating. By night fall, the northern end of the tunnel was in the Division's hands. Despite the failure to achieve the first day's objectives, the 3rd Division had helped to make an irreparable breach in the Hindenburg Line, denying the enemy the chance to rebuild its forces behind the strongest defensive system on the Western Front.[89]

On 2 October, Monash relieved the 3rd Division and British units took over the battle. The struggle had exhausted Gellibrand's infantry and greatly reduced the Division's strength. When the 44th Battalion withdrew it had only 80 men left.[90]

The war was now over for most of the 3rd Division and the rest of the Australian Corps. Only the artillery continued the advance, supporting other formations in their pursuit. Wherever the Germans attempted to make a stand, the British attacked and drove them back. On 17 October, the 3rd Division's 27th Battery supported an American attack near St Souplet and then moved forward in preparation for the next assault, which took place on 23 October. November saw the battery near Wassigny, where, on the morning of 4 November, its gunners fired their last barrage of the war. A few days later, on 11 November, Germany conceded defeat and agreed to an Armistice.[91]

For these men, who were really only civilians in uniform, the end of the war ushered in a welcome period of calm, but also of frustration at military authority and a longing to get home. The 41st Battalion was marching through Airaines when it received news of the Armistice. The city's inhabitants had decorated the streets with tricolour bunting and the church bells announced the German capitulation. Back at camp, the men began to celebrate, and on Sunday the battalion attended a victory mass. Some men broke camp, reasoning that their job was done, and a few got as far as Paris. For the rest, military discipline no longer seemed reasonable and the battalion en masse refused to do drill. Instead, the men went on route marches, played soccer or pelted one another with snow balls.[92]

In the 3rd Division's artillery units, the situation was more serious. The officers soon had a full scale mutiny on their hands. On 12 December, the entire 7th Field Artillery Brigade refused to turn out for a march and, in early January 1919, the commander of the 27th Battery placed all his men under arrest. With the war over, the men no longer accepted the constraints of being a soldier, and desired greater access to rest and recreation. The officers were in part to blame. Used to the level of obedience during wartime, they overreacted and several lost their commands as a result. Tensions eased when the divisional artillery moved to a larger town with better facilities for the men's entertainment, including a cinema.[93]

In March 1919, the pace of demobilisation accelerated. As shipping became available, 1000-man cohorts of troops, the capacity of a standard troopship, left for the long voyage home. One by one, the Army disbanded the Division's formations and, by early May, barely 1400 troops remained in France. On 9 April, Gellibrand handed his command over to Brigadier-General James H. Cannan, the 11th Brigade's former commander. It was Cannan who oversaw the Division's final days before its disbandment. The final entry in the 3rd Division's War Diary on 28 May, was a report which contained the announcement that the foundation for the formation's monument at Mericourt L'Abbaye had been finished.[94]

For those who had survived, the great adventure was over. The men of the 3rd Division had done their duty and done it well. Two years earlier, when they reported to training camps across Australia, they were green troops, late to the game, who deserved the derision of the senior divisions. By the war's end, there was no doubt that the 3rd Division was a veteran formation which had met the enemy at its strongest and been victorious. In 1919, the troops returned home, removed their uniforms and resumed their civilian lives. Although no longer soldiers, they left an enduring legacy in what they had accomplished.

Endnotes

1 M. B. B. Keatinge, *War Book of the Third Pioneer Battalion* (Melbourne: The Specialty Press Pty Ltd, 1922), p. 12; Eric Fairey, *The Story and Official History of the 38th Battalion, AIF* (Bendigo: Cambridge Press, 1920), p. 1; Anon., *The Thirty-Ninth: The History of the 39th Battalion Australian Imperial Force* (Melbourne: G. W. Green & Sons Pty. Ltd., 1934), p. 30; Anon., *The History of the Ninth Field Ambulance* (Sydney: John Sands, Ltd., 1919), p. 13; and Anon., *Souvenir, Being an Unofficial Résumé of the History of the 11th Australian Field Ambulance* (Eleventh Australian Field Ambulance, 1919), p. 12.

2 'CRA 3rd Division to C. B. White', 26 July 1916, in 3rd Division Artillery Reports, AWM224, Mss 4; Fairey, *The Story and Official History of the 38th Battalion, AIF*, p. 1; and Keatinge, *War Book of the Third Pioneer Battalion*, pp. 21-23.

3 E. J. Colliver and B. H. Richardson, *The Forty-Third: The Story and Official History of the 43rd Battalion, AIF* (Adelaide: Rigby, 1920), pp. 9-12; and H. G. McNicol, *The Thirty-Seventh: A History of the Thirty-Seventh Battalion AIF* (Melbourne: Modern Printing Co. Pty Ltd, 1936), pp. 5-6.

4 Mary Reddrop, ed., *Jim's Story: With the 37th Battalion – AIF* (Melbourne: Spectrum Publications, 1982), pp. 8-9; Fairey, *The Story and Official History of the 38th Battalion, AIF*, p. 2; C. Longmore, *'Eggs-A-Cook': The Story of the Forty-Fourth, War as the Digger Fought It* (Perth: The Colortype Press Limited, 1920), pp. 8-10; and Colliver and Richardson, *The Forty-Third*, pp. 12-13.

5 Diary, Schwinghammer Collection, 2DRL234, AWM.

6 Colliver and Richardson, *The Forty-Third*, pp. 13-16; Keatinge, *War Book of the Third Pioneer Battalion*, pp. 24-25; and Anon, *Short History of the 34th Battalion, A. I. F.* (34th Battalion A.I.F. Association, nd), pp. 3-4.

7 'Letter of 6 June 1916', Rose Collection, AWM, 2DRL003.

8 Anon., *The Thirty-Ninth*, pp. 35-36.

9 'Letter of 6 June 1916', Rose Collection, AWM, 2DRL003; 'Letter to Sister Maggie', 25 June 1916, Bellingham Collection, AWM, PR91/046; and Reddrop, *Jim's Story*, p. 11.

10 Fairey, *The Story and Official History of the 38th Battalion, AIF*, pp. 3-4; and Diary, Schwinghammer Collection, 2DRL234, AWM.

11 Diary, Schwinghammer Collection, 2DRL234, AWM; 'Letter to Sister Maggie', 25 June 1916, Bellingham Collection, AWM, PR91/046; Anon., *The Thirty-Ninth*, p. 37; and Keatinge, *War Book of the Third Pioneer Battalion*, p. 25.

12 'Letter of C. H. Peters', 31 July 1916, Peters Collection, AWM, 1DRL545; and Longmore, *'Eggs-A-Cook': The Story of the Forty-Fourth*, p. 10.

13 Edwards, *Never a Backward Step*, p. 23.

14 'Letter to Sister Maggie', 15 July 1916, Bellingham Collection, AWM, PR91/046; and Anon., *The Thirty-Ninth*, p. 38.

15 'Letter of Cpl L. R. Boulton', 20 September 1916, Boulton Collection, AWM, 1DRL136; Anon, *The Thirty-Ninth*, pp. 42-43; McNicol, *The Thirty-Seventh*, p. 14; and Gammage, *The Broken Years*, p. 34.

16 Diary, Schwinghammer Collection, 2DRL234, AWM; 'Letter of Spencer E. Jenkins', 24 July 1916, Author's Possession; Fairey, *The Story and Official History of the 38th Battalion, AIF*, p. 5; and Colliver and Richardson, *The Forty-Third*, p. 15.

17 'Letter of Spencer E. Jenkins', 24 July 1917 (author's possession).

18 P. A. Pedersen, *Monash as Military Commander* (Melbourne: Melbourne University Press, 1985), pp. 142-44.

19 *Ibid.*, p. 145.

20 Geoffrey Serle, *John Monash: A Biography* (Melbourne: Melbourne University Press, 1982), p. 269.

21 *Ibid.*; Pedersen, *Monash as Military Commander*, p. 146; F. M. Cutlack, ed., *The War Letters of General Monash* (Sydney: Angus and Robertson, 1935), p. 129; Anon., *The Thirty-Ninth*, p. 54; and 'CRA 3rd Division to C. B. White', 26 July 1916, in Third Divisional Artillery Reports, AWM224, Mss 4.

22 Anon., *The Thirty-Ninth*, pp. 54-55; and 'Letter to Sister Maggie', 6 October 1916, Bellingham Collection, AWM, PR91/046.

23 *History, 10th Field Company, Engineers* (1919), p. 10; and Fairey, *The Story and Official History of the 38th Battalion, AIF*, p. 7.

24 'Report on the Third Australian Division by Major General Francis Howard, Inspector of Infanty', 27 August 1916, AWM, Monash Collection, 3DRL2316, item 22.

25 'War Office to GOC Southern Command' 18 August 1916, 'Monash to HQ Southern Command' 21 August 1916, and 'GOC AIF Depots in UK to HQ Southern Command', all in Monash Collection, 3DRL2316, AWM. *See also* C. E. W. Bean, *The Australian Imperial Force in France, 1916* (Sydney: Angus and Robertson, 1929), pp. 866-67; and C. E. W. Bean, *The Australian Imperial Force in France During the Main German Offensive, 1918* (Sydney: Angus and Robertson, 1937), p. 3.

26 'Letter to Massie', 27 September 1916, Crowley Collection, AWM, PR00618; 'Letter to Rob', 18 April 1917, L. W. Robinson Collection, AWM, 3DRL2147; and Anon., *The Thirty-Ninth*, pp. 57-58.

27 Information taken from 'Units and Formations of the AIF Serving Abroad', AA(VIC), MP367/1, item 469/2/418.

28 Reddrop, *Jim's Story*, p. 58; and F. C. Green, *The Fortieth: A Record of the 40th Battalion, AIF* (Hobart: Government Printer, 1922), p. 5.

29 Fairey, *The Story and Official History of the 38th Battalion, AIF*, pp. 7-8; Green, *The Fortieth*, p. 6; McNicol, *The Thirty-Seventh*, p. 31; Anon., *The Thirty-Ninth*, pp. 66-67; Reddrop, *Jim's Story*, p. 59; and Cuttriss, *'Over the Top' With the Third Australian Division*, p. 21.

30 Fairey, *The Story and Official History of the 38th Battalion*, p. 8; and Anon., *The History of the Ninth Field Ambulance*, p. 16.

31 Anon., *The History of the Ninth Field Ambulance*, p. 16.

32 Reddrop, *Jim's Story*, pp. 66-67. The *minniwerfer* was a German trench mortar bomb which weighed 190 pounds.

33 'Relief Table, 9th Australian Infantry Brigade', in AWM4, 34th Battalion War Diary. *See also* War Diary entries for the 33rd and 34th Battalions, also in AWM4.

34 Fairey, *The Story and Official History of the 38th Battalion*, pp. 10-11.

35 'War Diary – 10 Field Company', 27 November to 16 December 1916, AWM4; Anon., *The History of the Ninth Field Ambulance*, p. 16; *Souvenir, Being an Unofficial Résumé of the History of the 11th Australian Field Ambulance*, p. 12.

36 'Letters of EAC Atkinson, 8 December 1916 and 21 January 1917, AWM, 2DRL59; and 'Letter of Bellingham', 1 February 1917, PR91/046, AWM.

37 'War Diary – 33rd Battalion', and 'War Diary – 34th Battalion', 29 November 1916, AWM4.

38 Reddrop, *Jim's Story*, pp. 66-67.

39 Anon., *The Thirty-Ninth*, p. 79.

40 C. E. W. Bean, *The Australian Imperial Force in France, 1917* (Sydney: Angus and Robertson, 1943), p. 563; and Fairey, *The Story and Official History of the 38th Battalion*, p. 11. Garland received the Military Cross and Meyerink the Military Medal. *See* 'Honours and Awards, 3rd Division, 7 December 1916 to 22 December 1916, AWM28.

41 Anon., *The Thirty-Ninth*, p. 79.

42 Longmore, *'Eggs-A-Cook': The Story of the Forty-Fourth*, pp. 26-27; and Anon., *The Thirty-Ninth*, pp. 84-85.

43 Bean, *The Australian Imperial Force in France, 1917*, p. 565.

44 Reddrop, *Jim's Story*, pp. 103-5; McNicol, *The Thirty-Seventh*, pp. 62-65; and Fairey, *The Story and Official History of the 38th Battalion, AIF*, pp. 17-18. *See also*, C. H. Peters, 'Raiding Days', AWM MS1255.

45 Bean, *The Australian Imperial Force in France, 1917*, p. 568 and Longmore, *'Eggs-A-Cook': The Story of the Forty-Fourth*, p. 27.

46 Anon., *The Thirty-Ninth*, pp. 103-5.

47 Longmore, *'Eggs-A-Cook' : The Story of the Forty-Fourth*, pp. 41-42.

48 *History, 10th Field Company, Engineers*, p. 19; Keatinge, *War Book of the Third Pioneer Battalion*, pp. 52-54; and Anon., *The Thirty-Ninth*, pp. 114-15.

49 *Short History of the 34th Battalion, A.I.F.*, p. 15.

50 Keatinge, *War Book of the Third Pioneer Battalion*, p. 54.

51 Fairey, *The Story and Official History of the 38th Battalion*, pp. 24-25.

52 Bean, *The Australian Imperial Force in France, 1917*, pp. 712-21.
53 'Letter of W. J. Rose', 4 October 1917, Rose Collection, AWM, 2DRL1003.
54 Diary, Schwinghammer Collection, 2DRL234, AWM.
55 Colliver and Richardson, *The Forty-Third*, pp. 53-59; Longmore, *'Eggs-A-Cook': The Story of the Forty-Fourth*, pp. 64-66; Green, *The Fortieth*, pp. 74-78; McNicol, *The Thirty-Seventh*, pp. 128-33; and Keatinge, *War Book of the Third Pioneer Battalion*, p. 188.
56 Bean, *The Australian Imperial Force in France, 1917*, p. 871-72.
57 Green, *The Fortieth*, p. 78.
58 'Letter of Spencer E. Jenkins, 9 October 1917 (author's possession).
59 Cutlack, *War Letters of General Monash*, p. 197.
60 Bean, *The Australian Imperial Force in France, 1917*, p. 883.
61 *Ibid.*, p. 906.
62 'Letter of G. M. Carson', 25 October 1917, Carson Collection, AWM, 2DRL185; Edwards, *Never a Backward Step*, p. 52; and Anon., *The Thirty-Ninth*, pp. 152-53.
63 Anon., *The History of the Ninth Field Ambulance*, p. 28.
64 *Ibid.*, p. 30.
65 'Letter of J. Hardie', 30 October 1917, Hardie Collection, AWM, PR00519.
66 'Letter of G. M. Carson', 25 October 1917, G. M. Carson Collection, AWM, 2DRL185; and Anon., *The Thirty-Ninth*, pp. 152-56.
67 Anon., *The Thirty-Ninth*, pp. 152-56; Cutlack, ed., *War Letters of General Monash*, p. 200.
68 John Monash, *The Australian Victories in France in 1918* (London: Hutchinson, 1920), p. 28.
69 Cutlack, ed., *War Letters of General Monash*, pp. 220 and 227-8; Bean, *The Australian Imperial Force in France, 1918*, pp. 153-54.
70 Fairey, *The Story and Official History of the 38th Battalion*, p. 48; Keatinge, *War Book of the Third Pioneer Battalion*, pp. 85-86.
71 Diary, Schwinghammer Collection, AWM 2DRL234.
72 Longmore, *'Eggs-A-Cook': The Story of the Forty-Fourth*, pp. 88-89.
73 T. D. Bridger, *With the 27th Battery in France* (London: St Clements Press, 1919), pp. 91-92; and Fairey, *The Story and Official History of the 38th Battalion*, pp. 49-50.
74 *History 10th Field Company*, p. 37; and *History of the 11th Field Company*, pp. 22-23.
75 Longmore, *'Eggs-A-Cook' The Story of the Forty-Fourth*, pp. 83-84; Diary, Schwinghammer Collection, AWM 2DRL234.
76 'Daylight Raid, HQ 33rd Battalion to HQ 9th Infantry Brigade', 21 June 1918, AWM26, item 394/5.
77 'Murray to His Father', 28 August 1918, Murray Collection, PR00594, AWM; and 'Reflections on the Push of 7-8 August', J. S. Finney Collection, AWM, PR00266.
78 'Reflections on the Push of 7-8 August', J. S. Finney Collection, AWM, PR00266.
79 Keatinge, *War Book of the Third Pioneer Battalion*, pp. 114-15.
80 Longmore, *'Eggs-A-Cook': The Story of the Forty-Fourth*, p. 112; and 'Murray to His Father', 28 August 1918, Murray Collection, PR00594, AWM.
81 Monash, *The Australian Victories in France in 1918*, p. 158; and Robin Prior and Trevor Wilson, *Command on the Western Front: The Military Career of Sir Henry Rawlinson, 1914-18* (Oxford: Blackwell, 1992), p. 337.
82 Keatinge, *War Book of the Third Pioneer Battalion*, pp. 125-34.
83 Diary, Schwinghammer Collection, 2DRL234, AWM.
84 Vivian Brahms, *The Spirit of the Forty-Second: Narrative of the 42nd Battalion, 11th Infantry Brigade, 3rd Division, Australian Imperial Forces during the Great War, 1914-1918* (Brisbane: W. R. Smith & Paterson, 1938), pp. 105-9; and 'Action at Bony', Gellibrand Collection, 3DRL1473, item 103, AWM.
85 Prior and Wilson, *Command on the Western Front*, pp. 346-48.
86 'Operations of the 3rd Aust. Division in Conjunction with the 27th American Division in Operations Against the Hindenburg Line 29/9 – 2/10/1918', Gellibrand Collection, 3DRL1473, item 103, AWM.
87 'Phone Message from Lieut. Phillips – 10th Brigade', 29 September 1918, Gellibrand Collection, 3DRL1473, item 103, AWM.

88 'Remarks on the Operation', Gellibrand Collection, 3DRL1473, item 103, AWM.
89 Monash, *The Australian Victories in France in 1918*, pp. 267-68.
90 Longmore, *'Eggs-A-Cook': The Story of the Forty-Fourth*, p. 123.
91 Bridger, *With the 27th Battery in France*, pp. 162-66.
92 Diary, Schwinghammer Collection, 2DRL234, AWM.
93 'Lieut. Col. Alsop, CO 3rd Division Artillery to GOC 3rd Division', 12 December 1918, Gellibrand Collection, 3DRL1473, item 67, AWM. *See also*, 'Lieut. Col. Alsop, CO 3rd Division Artillery to HQ Australian Corps', 29 December 1918 and 5 January 1919; and 'Grimwade to Gellibrand', 17 January 1919 all in Gellibrand Collection, 3DRL1473, item 58, AWM.
94 'Strength Return' 3 May 1919, 'Special Order', 10 April 1919, and 'Final Report of Division Group' 28 May 1919, all in AWM4, item 1/47. *See also*, Peter S. Sadler, *The Paladin: A Life of Major-General Sir John Gellibrand* (Melbourne: Oxford University Press, 2000), pp. 187-88.

Chapter 3
The 3rd Australian Division Between the World Wars

Post-War Reorganisation

Upon the AIF's disbandment, Australia's military power again reverted to its home forces, the Citizen Forces' of the Kitchener scheme which the government had begun to raise in 1911. When the war began in 1914 the Army had implemented only about 60 per cent of the planned organisation. After the declaration of war, Australia virtually suspended the Home Army's further development and instead focused its efforts on the raising and maintenance of the AIF. By early 1915, it was clear that the country was unable to support two armies. For obvious reasons, the AIF became the priority and the Citizen Forces began to decay. Corresponding to the AIF's progressive expansion and gain in stature was the Home Army's decline, as its best officers and men volunteered for service overseas.

In September 1915, the Military Board suspended training of all non-AIF units. Soon after, it also stopped issuing units with uniforms or equipment. The only significant training the Citizen Forces received during the World War I was in 1917, when the Military Board temporarily resumed home training and authorised an eight-day camp of continuous instruction. However, the shortage of qualified teaching staff and deficiencies in equipment made this effort meaningless. By 1918, the diversion of officers, soldiers, equipment and resources to the AIF meant that most Citizen Forces units existed little more than on paper. In June 1918, Pearce, the Minister for Defence, admitted the extent of the problem and announced that the 'condition of the Citizen Forces has been giving the government concern for some considerable time, owing to the fact that many units have approached the vanishing point'. He continued that the Citizen Forces was now composed of 'youths insufficiently officered and necessarily inadequately trained'.[1] By the time of the Armistice in 1918, the army Australia had begun to raise in 1911 was, for all realistic intents, gone.

In 1919, Australia's military capability consisted of a pool of AIF veterans, now returned to their civilian occupations, and the moribund Citizen Forces. The Permanent Forces were also available but their purpose remained the administration and training of the Citizen Forces, the maintenance of coastal defences and the provision of technical services. Most regular soldiers served as staff officers or instructors in Citizen Forces units, and the Permanent Forces had no field force capability of their own.[2] In fact, the condition of the army was such that the Military Board controlled a force which existed in name only and which was incapable of conducting any organised or effective operation. In the aftermath of victory over Germany, Australia was defenceless. Its military leaders had to return to the beginning and craft a new army for the nation's security. A new start, however, also represented a new opportunity. The AIF's senior officers returning from France and the Middle East arrived home with a vastly improved knowledge of, and tremendous practical experience in, waging modern war. The task of forming the new Australian Army fell to these highly capable veterans.

In January 1919, Pearce appointed a committee of senior officers to undertake the design of the post-war Army. As chairman, Pearce chose Lieutenant-General Sir Henry G. Chauvel. Also serving on the committee were Lieutenant-Generals Sir John Monash and Sir Cyril B. B. White, and Major-Generals Sir James W. McCay, Gordon Legge and Sir Talbot Hobbs. The Minister for Defence instructed the committee to address the requirements of a new force, including issues such as organisation, fixed defences, citizen and permanent forces, reserve forces and rifle clubs, amongst other matters. Pearce gave the officers a wide brief and only cautioned that their proposals had to be fiscally prudent. He reminded the committee that it was not to be too lavish in its recommendations, as the war had strained the nation's finances, and any scheme it proposed had to be within reason.[3]

The committee recognised that the Kitchener scheme had many deficiencies, especially in relation to the requirements for waging a conflict against a first-class opponent. The Kitchener organisation, in fact, was not an army in any operational sense but a collection of battalions, squadrons, batteries and field companies whose primary functions were either to deter or defeat a raiding force, or to serve in a coalition led by British officers in the case of a greater threat. It lacked all the modern features of military power whose necessity now appeared so obvious to those who had served with the AIF during World War I. The senior officers knew that armies require a vast array of divisional, corps and support units, in addition to many specialist elements, in order to maximise a force's combat power. The Australian Army Kitchener had designed lacked a divisional structure, as well as these critical support arms, and was incapable of sustaining itself against a comparable armed opponent for any length of time.

Under Chauvel's guidance, the committee produced a detailed report outlining the forces Australia required for its defence. The members based their recommendations on the principle that, while the nation would continue to remain dependent upon Britain for its ultimate security, it required a force of sufficient size to guard its vast coastline, to prevent an invasion or (if one did occur) to postpone a decisive decision until the arrival of assistance from the Empire. The army's leaders were also in no doubt as to their future opponent. Japan, they believed, posed the 'only potential and probable enemy'.[4] For the next twenty years, a series of army leaders adhered unshakeably to the opinion that Japan was Australia's most dangerous threat.

The committee met in January 1920 and Chauvel presented its findings to Pearce the following month. The officers recommended an army of 180,000 citizen soldiers as the minimum size to give the nation a chance of holding out until help arrived. The committee decided upon a force of four infantry divisions, two cavalry divisions (mounted infantry) and three mixed brigades, which could unite to form a fifth division. They also identified numerous non-divisional support units, such as tank battalions, heavy artillery batteries and chemical warfare companies. Their manpower estimate also included the troops needed to staff the coastal fortresses, as well as the members of the Permanent Forces. Not surprisingly, the army the committee recommended was similar in size and structure to the force that Australia had maintained on the Western Front.[5]

The Council of Defence approved the organisation in April and forwarded the proposal to Prime Minister William Hughes. In a related recommendation, the Council also supported the committee's suggestion that the government continue the existing Universal Training Scheme but with an increased training obligation that was to include a total of thirteen weeks of continuous instruction. The senior officers looked back on the extensive training the AIF had required in Egypt prior to its commitment at Gallipoli. Monash's experience with the 3rd Division further underscored the need for better preparation. When the formation began its training in England, it had undergone a lengthy work-up before being ready for the trenches. The senior officers understood that more training would allow units to achieve a higher standard of readiness – essential, they recognised, if the army was to accelerate its mobilisation and field an effective field force as quickly as possible. The Council of Defence also passed on to the government the committee's request for an adequate supply of guns, rifles, ammunition and other stores, including reserves. The army wanted to have sufficient stores on hand in order to provide realistic training and to arm any new units it needed to raise during mobilisation. The Council of Defence opined that the minimum operating budget for defence was £8.25 million a year, of which the army estimated its share as £3.5 million. This did not include any monies for capital works, such as the modernisation of the coastal forts.[6]

When Hughes took up the matter, he approved the broad outline of the senior officers' committee but baulked at its financial implications. Instead of an operating vote of £3.5 million, the army received considerably less. For all its expenses, including capital improvements, the army would have to make do with between £3 and £3.2 million. From that reduced figure, the Military Board then had to allocate approximately £2.4 million for capital works, equipment purchases, loan repayments and the overhead and administrative costs of the Permanent Forces. This left barely £650,000 for the organisation and training of the Citizen Forces.[7]

As a result of the reduced allocation, White, now serving as Chief of the General Staff (CGS), reduced the army's training objectives for the 1920/21 fiscal year. Instead of the desired thirteen weeks training, the army could not manage to maintain even the inadequate pre-war obligation of sixteen days training, divided between an eight-day camp of continuous instruction and eight days of home training. Instead, the Hughes budget permitted just eight days of continuous instruction, plus four days home training, and for only a single intake, the class of 1902. For the 1899–1901 intakes the army had funds sufficient for just four days' home training, while soldiers from earlier years had to report for only a single half-day parade.[8]

Moreover, the lack of funds meant that the army could not maintain its desired establishment of 180,000 troops and instead had to accept a smaller Citizen Forces of 118,000 and a Permanent Forces of 3500.

White and the other senior officers objected to the government's paltry funding, but without success. They argued that the funds allocated permitted little more than nominal training and weapons purchases and left the force insufficiently armed and equipped. The Military Board explained to Pearce that within the estimate there was practically no provision for the training of the rank and file, and that the army would accomplish little more than the implementation of the organisation's shell. They pointed out that the vote did not even allow the provision of ammunition to artillery units for practice firing. Pearce's reply was brief. He commented to the Military Board's members that their views would receive his full consideration. Shortly thereafter, the government insisted on a policy of defence economy.[9]

Hughes' determination to limit defence expenditure established an enduring pattern of governmental under-funding of military requirements that remained in place for the entire inter-war period. Instead of supporting the forces with adequate funds, successive governments preferred to purchase national security as cheaply as possible. As the course of the inter-war period would demonstrate, the Treasury's singular emphasis on economy prevented the army from creating an effective military instrument. The senior officers did succeed in creating the organisation on paper but it was to remain on a skeletal basis and lacked the resources to turn raw recruits into effective soldiers. In the end, consistent government refusal to consider properly the requirements of defence failed to provide the nation with the forces needed for its security.

In addition, the funding pattern Hughes established underscored the extent of the gulf which existed between the nation's professional soldiers and their political masters regarding the requirements and importance of national security as a function of government policy. By reducing the defence vote so radically, Hughes signalled his intention to rely on Britain and the Royal Navy as the primary bastions of Australian security. Furthermore, it also meant that the government had refused to take seriously the threat of Japan which the senior officers had identified. The army became a low priority in the nation's defence hierarchy and the government effectively limited its role to raid deterrence rather than the opposition of more serious threats. Unfortunately, Hughes' decision was not atypical but rather the beginning of a consistent policy that lasted until mobilisation commenced for World War II. Instead of resources improving, within two years the army's vote had declined further and did not recover until the onset of war.

The 3rd Division is Reborn

By the end of 1920, Army Headquarters had completed the allocation of units to formations. However, the new divisional structure did not come into effect until 1 May 1921. On that date, the 3rd Division officially rejoined the army's order of battle. Following

Table 2
Order of Battle, 3rd Division, 1921[10]

TYPE	FORMATION	UNIT
Headquarters		3rd Division Headquarters
Artillery		Headquarters, 3rd Division Artillery
		3rd Division Ammunition Column
	2nd Fd Bde	Headquarters, 2nd Field Brigade
		4th, 5th, 6th Batteries
		102nd (howitzer) Battery
	4th Fd Bde	Headquarters, 4th Field Brigade
		10th, 11th, 12th Batteries
		104th (howitzer) Battery
	8th Fd Bde	Headquarters, 8th Field Brigade
		29th, 30th, 31st Batteries
		108th (howitzer) Battery
Engineers		Headquarters, 3rd Division Engineers
		3rd Field Park Company
		4th, 10th, 15th Field Companies
Signals		3rd Division Signals
Infantry	4th Inf Bde	Headquarters, 4th Infantry Brigade
		14th Bn – The Prahran Regt
		22nd Bn – The Richmond Regt
		29th Bn – The East Melbourne Regt
		46th Bn – The Brighton Rifles
	10th Inf Bde	Headquarters, 10th Infantry Brigade
		24th Bn – The Kooyong Regt
		37th Bn – The Henty Regt
		39th Bn – The Hawthorn-Kew Regt
		48th Bn – The Torrens Regt
	15th Inf Bde	Headquarters, 15th Infantry Brigade
		57th Bn – The Merri Regt
		58th Bn – The Essendon Rifles
		59th Bn – The Coburg-Brunswick Regt
		60th Bn – The Heidelberg Regt
Supply & Transport		3rd Division Australian Army Service Corps
Medical		4th, 10th, 15th Field Ambulances
		3rd Field Hygiene Section
Veterinary		3rd Mobile Veterinary Section
Ordnance		3rd Division Ordnance*
Provost		3rd Provost Company*
Employment		3rd Divisional Employment Company*

note: * not raised in peacetime

pre-war practice, the army raised its units on a regional basis. To facilitate recruitment and training, the Military Board revised the Kitchener scheme's pre-war training organisation and divided the country into fifteen brigade areas, 60 battalion areas and 144 training areas. Each battalion, regiment, battery and other unit had its own recruitment area from which to draw its members. However, unlike in World War I, when the AIF had raised the Division from across the entire nation, the new 3rd drew largely upon Victoria, strengthing its association with that State. Table 2 outlines the 3rd Division's order of battle.

The structure outlined in 1920 remained in effect up until World War II. The only significant modification for the 3rd Division was the substitution of the 52nd Battalion (The Gippsland Regiment) for the 48th Battalion in the 10th Infantry Brigade. The 48th Battalion then joined the 4th Infantry Division as a part of its 3rd Infantry Brigade.

The assignment of particular numbers to the battalions of the 3rd Division came about from a decision Pearce had made in early 1917. Anticipating the end of the conflict, he concluded that it would be beneficial for the development of the post-war army if it could lay claim to the spirit and tradition of the AIF. Consequently, he decided that when the army returned to peace, its units would where practical carry the designation of the AIF unit which had recruited from the same catchment. While desirable, it was a highly artificial attempt to transfer battle honours from a disbanded organisation to its successor, and it created new problems. For the 3rd Division, it resulted in an odd situation in which only two of the formation's battalions – the 37th and 39th – carried the same number as those which had fought on the Western Front. In 1916, the army had raised the 3rd Division on a national basis, and its units came from all the States. As Victoria was now the 3rd Division's home, most of its battalions bore the designation of units which had actually served in other AIF divisions.[11]

On 8 February 1921, Gellibrand received the 3rd Division's command, renewing his association with the formation. However, his second tenure proved relatively brief and unremarkable. A little more than a year later, on 20 February 1922, he resigned his position to return to his home state of Tasmania.[12] He was succeeded by Major-General G. J. Johnston.

With an establishment of only 118,000, the army could not staff its units at their full strength. Instead, the Military Board set a lower peace-time establishment. Units would only reach their war-time strength on mobilisation for war. For the other ranks the Military Board decided to maintain the infantry, artillery, service, signal and ordnance units at 90 per cent of war establishment, and cavalry, engineers, medical and veterinary units at 60 per cent of war establishment. All units were to maintain officer levels at full strength. This meant that under peace time conditions, the 3rd Division's establishment was 16,000 officers and other ranks. Only on mobilisation would the formation expand to its fighting strength of 18,400.[13]

Under the Universal Training Scheme, the army drew upon men from intakes of 1896 to 1902 and allocated them amongst Citizen Force units. Most of the 3rd Division's units quickly achieved their desired strength. In fact, some had more men posted than there was room on their establishment. For example, the 29th Battalion was a massive unit, with 1445 officers and other ranks, and the 59th Battalion was only one soldier smaller. Only three battalions numbered less than a thousand, the smallest being the 57th with 702 members. In addition to conscripts, the 3rd Division's ranks included a small number of volunteers but these numbered less than 150.[14]

At first the development of the 3rd Division, as with the rest of the army, proceeded along the lines which Chauvel and the other senior officers had outlined. However, that situation did not last long. In fact, 1921 proved the high point of the inter-war army, and its designers' ambitious plans soon lay in ruins. Instead of the gradual development of an effective military force, instilled with the experience and pride so hard won during World War I, the army quickly sank into irrelevance, inefficiency and near oblivion. For fiscal year 1921/22, the Hughes government authorised an army vote similar to the previous one. However, the government's defence policy was about to change dramatically, and the result was savage and sustained cuts to the army's already inadequate estimate. The onset of the Great Depression in 1929 led to even greater calls for economy from the services and only exacerbated an already grim situation. The ramifications for the training and arming of the 3rd Division were serious indeed.

In a politely worded statement in his Inspector-General report of May 1922, Chauvel pointed out that:

> Owing to the constant changes and curtailment in the approved programmes of training necessitated by financial considerations, it cannot be said that, except in organization and the training of officers and non-commissioned officers, any progress has been made during the past year.[15]

In writing these words, Chauvel could not have known that the situation would grow worse. As World War II approached, the condition of the army was dire and only a belated mobilisation in 1938 could begin to undo years of neglect. By then, of course, it was too late, and the nation's military forces entered that conflict unprepared to wage war. While these decisions occurred at the highest level of government policy and army administration, their effects were felt throughout the ranks of the 3rd Division.

The first turning point in the fortunes of the Citizen Forces occurred in 1922, when representatives of a number of countries meeting in Washington reached an agreement on how to respect the integrity of China and secure peace in the Pacific. From the Australian perspective, the key treaty was the one on naval disarmament that was to become known as the Washington Naval Treaty. Its signatories – Britain, the United States, Japan, France and Italy – agreed to tonnage limits on the capital ships in their fleets. For example, the United States and Britain could possess 500,000 tons of capital ships, Japan 300,000 and France and Italy 175,000 each. As a result of this accord, the Hughes government assumed that Japan no longer posed a threat. The new Minister for Defence, Walter Massey-Greene, immediately halved the army's budget. In fiscal year 1921-22, the army's allocation had been £2.2 million. For 1922-23, it received less than £1.1 million. While the army's share of defence monies did improve somewhat during the decade, it never exceeded about £1.8 million.

The following year, Britain held an Imperial Conference in London. For Australia, this conference was the most critical meeting of the inter-war period, as it resulted in Britain's agreement to build a major naval base at Singapore. In the case of war with Japan, the Royal Navy would deploy to the east and base itself at Singapore in order to deter or defeat the enemy's fleet. The Singapore Strategy, as it became known, became the cornerstone of Australian security policy for the remainder of the inter-war period and on into World War II, until that base's capture by the Japanese in 1942.[16]

The government's acceptance of the Singapore Strategy greatly affected the role of the army in the nation's defence and had serious consequences for the organisation of the 3rd Division. When the senior officers designed the army's organisation, they had accepted that they needed a large, reasonably well trained force which could hold off an invader until the Empire mobilised its strength. After the signing of the Washington Naval Treaty, and in the light of Britain's commitment to Singapore, the Australian government no longer saw a need to maintain a significant ground force. The government concluded that Australia's defence was no longer an Australian responsibility but instead a matter for London and the Royal Navy. The only task for which the army had to prepare was that of repelling small landing parties from an isolated raider that managed to slip past the fleet. This abrogation of duty allowed the government the opportunity to pursue its preferred options on security policy, namely the containment of costs and the imposition of defence economies.

For fiscal year 1921-22, the army had mustered Citizen Forces numbering 118,400 and Permanent Forces totalling 2495. By the following year these figures had declined to 31,000 and 1600 respectively.[17] The army had planned to staff the 3rd Division at approximately 90 per cent of war establishment. Now the formation had to make do with barely 25 per cent of its required troops. Moreover, the limited funds cut the amount of training that was fiscally possible. Johnston's training program would have to make do with camps of just six days, supplemented by four days' home practice. In addition, the Military Board was able to afford training for only two quotas of Universal Trainees, those of eighteen and nineteen years of age, which further reduced the number of soldiers in the Division's active ranks.[18]

As a result of the reduction to 31,000, the army's leaders no longer expected the Citizen Forces to serve as a field force. Instead, their primary objective became that of training leaders. The Military Board employed the term 'nuclear structure' to describe the new policy. By 'nuclear' they meant that the Citizen Forces, as presently constituted, represented the minimum core force essential for the organisation's survival and the training of a new generation of leaders. Eventually, they hoped, the army would re-expand to the strength the senior officers had envisioned, using the truncated base as a guide. Moreover, by focusing on officer and non-commissioned officer training the army would have many of the leaders it would require when expansion took place. In effect, the nuclear rationale allowed the army to survive in a mothballed state. By training leaders, the force sustained an operational potential, but then required a lengthy rebuilding before it was capable of taking the field.

For the rest of the inter-war period, the 3rd Division had to grapple with the implications of the defence funding collapse and the reduction in manpower. Establishments and money became the critical determinants of the Division's effectiveness for the next seventeen years. How the 3rd Division negotiated these problems is the subject of the remainder of this chapter.

Manpower

Surviving documentation of the 3rd Division's manpower levels during the inter-war years is incomplete, but sufficient records exist to show the great difficulty its units faced in reaching their establishments and maintaining personnel. Although the government did gradually increase the Citizen Force's manpower ceiling, by 1929, it stood at only 49,000, still well below the senior officers' goal.[19] However, during the 1930s, these numbers again plunged. In 1930, the government reduced the army's establishment to 35,000 but the force did not reach even that modest goal until late in the inter-war period. As a result, the army remained a fragile force, organised around the nuclear concept, almost until the commencement of hostilities in 1939.

The wartime establishment of an infantry battalion was 75 officers and 1152 other ranks, divided between four companies and a headquarters wing. As a result of the government's change of policy direction in 1922, the army reduced a battalion's peacetime establishment to just 25 officers and 384 other ranks. Table 3 outlines the theoretical organisation of a battalion under the nuclear concept.

Table 3
Organisation of a Nuclear Battalion[20]

SUB-UNIT	NO. OFFICERS	NO. OTHER RANKS	TOTAL
HQ Wing	7	76	83
A Company	6	106	112
B Company	6	106	112
C Company	nil	nil	nil
D Support Company	6	96	102
Total	25	384	409

C Company simply did not exist, and the remaining companies typically had only two platoons rather than the normal four. The headquarters wing was supposed to have an anti-tank section but that was not raised.

Of course there were many exceptions to the outline above, depending on the ever-shifting manpower situation. For example, in 1926, the 46th Battalion's strength was so far below the required level that the Divisional Adjutant suggested its reduction to a two-company basis.[21] During the period, variations in organisation remained the hallmark of the force's structure. In 1934 the 24th and 37th/52nd Battalions consisted of four small companies each, plus a headquarters wing. In the same year, the 46th and 59th Battalions still operated on a three-company basis. Of its expected sixteen platoons, the 46th was missing five. At just seven platoons, the 59th Battalion was even weaker.[22]

After the installation of James Scullin as Prime Minister on 22 October 1929, the army's manpower situation dramatically worsened. Acting on the Labor Party's long-standing hostility to conscription, he summarily suspended the Universal Training Scheme soon after taking office. The government insisted that, instead of using conscription, the army staff its ranks through voluntary enlistment alone. To highlight the change, the Military Board abandoned the use of the term Citizen Forces for its non-permanent branch and instead designated them as the Militia. The force's strength declined rapidly as conscripts took their discharge and volunteers failed to come forth in sufficient numbers to make up the difference. By mid-1930, the army's citizen soldiers numbered approximately 23,700, well short of the 35,000 goal. It was to take the army until 1936 to reach its establishment.[23] The elimination of Universal Training thus had a major effect on the force's administration of manpower. Instead of a guaranteed annual intake, the army now had to compete for recruits.

The Great Depression was an extremely difficult period during which to maintain an army by voluntary means. The government's first priority was to save money, and it refused to offer appealing inducements to recruits. In fact, the terms of service were quite unattractive. The Scullin government expected volunteers to serve for only four shillings per day, the same rate conscripts had received under Universal Training. Moreover, drill halls and equipment were generally in poor repair, ammunition for live practice was limited, and training camps were repetitive or unimaginative owing to the need to curtail expenses. In fact, until 1936, the government even expected soldiers to pay for their own transportation to mandatory parades.[24]

The dire employment situation which existed during the Depression added further difficulties to the recruitment of soldiers. During a time of generally high unemployment and job uncertainty, such as the Great Depression, workers had even less security and more incentive not to antagonise their bosses. The result was that, in the 1930s, any soldier who reported to a camp of continuous instruction did so either with his employer's permission or risked his own and his family's livelihood. Unit commanders made frequent references to the lack of support by employers and the influence of job security in determining the ability of a soldier to meet his training obligations. For example, a report on the camps held by the 3rd Division's field companies in 1938 explained that only 64 per cent of the other ranks attended, 'due to the fear of losing employment or being victimised by the man's employer'.[25]

The promise of stable employment even influenced the location of training centres. The 37th/52nd Battalion raised two platoons in Yallourn because the State Electricity Commission was a major employer in the town. Yallourn thus offered the battalion a source of recruits who could count on stable employment. These soldiers were less likely to become non-effective, or need to search for work elsewhere.[26]

The adverse recruiting environment had a debilitating effect on the 3rd Division's manpower, and as a result the army had to reduce the formation's establishment levels even further. By 1931, only two battalions still had an establishment set at greater than 409 all ranks, and only one – the 37th/52nd Battalion – achieved this goal. In fact, Army Headquarters adopted a policy of variable establishments determined by local conditions. Table 4 draws upon strength returns for 1931 to contrast the official establishment and posted strength of the 3rd Division's battalions.

Table 4
Establishment and Strength of 3rd Division Battalions – 1931[27]

FORMATION	UNIT	ESTABLISHMENT	STRENGTH	STRENGTH TO ESTABLISHMENT PERCENTAGE
4th Inf Bde	14th Bn	295	282	96
	29th/22nd Bn	232	214	92
	46th Bn	245	240	98
10th Inf Bde	24th Bn	315	306	97
	37th/52nd Bn	443	409	92
	39th Bn	412	335	81
15th Inf Bde	57th/60th Bn	339	331	98
	58th Bn	281	277	99
	59th Bn	290	281	97

At that time, the 3rd Division's overall establishment was 4505 all ranks and its posted strength was 4221 or 95 per cent.[28]

Three years later, the manpower situation had become even worse as discharges outpaced enlistments and the Division experienced the full effect of the suspension of Universal Service. Table 5 summarises battalion strengths in 1934.

As table 5 shows, in 1934, only one battalion had more than 300 members, the average size being 243. The strongest battalion's strength did not reach 80 per cent of establishment, while the weakest achieved only 54 per cent of its approved size. These figures clearly demonstrate that, even with varied establishments, units were simply unable to recruit or retain men in sufficient numbers to maintain themselves at authorised strengths.

Table 5
Establishment and Strength of 3rd Division Battalions – 1934[29]

FORMATION	UNIT	ESTABLISHMENT	STRENGTH	STRENGTH TO ESTABLISHMENT PERCENTAGE
4th Inf Bde	14th Bn	280	195	70
	29th/22nd Bn	280	193	69
	46th Bn	384	275	72
10th Inf Bde	24th Bn	521	281	54
	37th/52nd Bn	521	325	62
	39th Bn	384	245	64
15th Inf Bde	57th/60th Bn	280	209	75
	58th Bn	366	239	65
	59th Bn	280	222	79

Tables 4 and 5 also show that the army had resorted to the expedient of linking battalions in order to stretch manpower resources. This first occurred in 1930, when each of the 3rd Division's infantry brigades linked two battalions and the formations henceforth operated on a three-battalion basis. On a war footing, the brigades would have an establishment of 3327 officers and other ranks. However, their peacetime establishments ranged from a low of 1005 for the 4th Infantry Brigade to 1213 for the 10th Infantry Brigade. The army set the 15th Brigade's establishment at 1109.[30]

After the suspension of Universal Service the artillery brigades also lost a battery and each headquarters now commanded just three batteries. The status of 2nd Field Artillery Brigade in 1931 demonstrated the need for this step. On mobilisation, the formation had an establishment of 876 officers and other ranks. Under the voluntary recruitment system, the army lowered its goal to 300, and in September, numbers stood at 288, or 96 per cent of its reduced authorisation. However, the formation was unable to sustain its numbers and, by 1935, its strength had sunk to 236, or 79 per cent of requirement. The other brigades were in a similar condition. With such limited manpower, the maintenance of four batteries per brigade was simply impossible. The surviving batteries were: 2nd Field Brigade (4th, 5th, 102nd Batteries), 4th Field Brigade (10th, 11th, 104th Batteries), and 8th Field Brigade (29th, 30th, 108th Batteries).[31]

The strength returns of the early 1930s also demonstrate the high rates of variation in the 3rd Division's official establishment. Table 6 outlines the speed with which it changed. The 3rd Division's attempts to maintain itself at a strength above 5000 (all ranks) were consistently unsuccessful. After March 1931, the formation's establishment pattern oscillated within a narrower range but, as late as 1937, its leaders made no further attempts to exceed the 5000 barrier.

Table 6
Variation in 3rd Division Establishments[32]

REPORTING PERIOD	ESTABLISHMENT STRENGTH	POSTED STRENGTH	POSTED/ESTABLISHMENT PERCENTAGE
September 1930	5,084	3,887	76
March 1931	5,160	4,399	85
June 1931	5,161	4,505	87
September 1931	4,505	4,221	95
December 1931	5,130	3,915	76
March 1932	4,507	4,254	94

The inter-war manpower crisis reached its most dire point when Major-General T. H. Dodds, the Adjutant-General, proposed a drastic reorganisation, the disbandment of several formations including the 3rd Division. He pointed out that by eliminating these organisations, the army could reduce its overhead and administration expenses and the savings could provide for additional militia training. Dodds first raised the issue in March 1931, in a letter to the then divisional commander, Major-General Harold Elliott.[33]

A year later, the proposal was sufficiently advanced for Dodds to send out a further warning letter. This time it was up to the 3rd Division's new commander, Major-General Thomas Blamey, to contemplate the formation's fate. Dodds advised of the imminent disbandment of not only the 3rd Division, but also the 4th Division as well as the 2nd Cavalry Division, the 3rd Military District Base and formations in other military districts. Dodds requested that the information be kept confidential, pending a decision and announcement by the Minister. Upon further reflection, however, Pearce, again serving as Minister for Defence, decided to retain the existing structure despite its great weaknesses.[34]

The failure to find enough men to staff the units properly threatened to undermine the entire rationale for the Militia's existence. The main objective of the nuclear army concept was to provide training opportunities for its leaders, both officers and non-commissioned officers. However, effective leadership preparation required units to have a pool of soldiers sufficient in size for the performance of real tasks and responsibilities through which leaders could develop their skills. In fact, the primary explanation for why the Military Board retained all these under strength units was that a large number of weak units provided more leadership positions than would a smaller number of full strength units. The comments of the Headquarters 15th Brigade highlighted this point. It argued against the disbandment of any company headquarters, as their elimination, despite being justified from a rational assessment of company strengths, would have had an adverse effect on the number of officers required by the organisation. Instead, 15th Brigade Headquarters suggested the elimination of platoons. The other ranks largely made up the platoon and its loss would not have as great an effect on the allocation of leadership positions as the disbandment of a company headquarters.[35]

As the figures presented in Table 6 show, the 3rd Division had barely enough soldiers to staff its order of battle and sustain its officer and non-commissioned officer positions for leadership training. This was a serious situation. As the army had abandoned the idea of creating a field force, the mandate of the nuclear organisation was its only justification for survival at these lowered levels. If it could not reach the leadership training goals of the nuclear organisation, it would provide the government with a tempting opportunity to reduce the force even further. In an era in which the government had already indicated overwhelming *ennui* for domestic defence issues, and demonstrated a misguided readiness to transfer its responsibility for national security to Imperial guarantees, the army could not take this threat idly.

Assessments of units conducted during the 1938 camps of continuous instruction demonstrated the difficulties the 3rd Division units had in meeting their nuclear organisation obligation. The key requirement was for a unit to have the ability to expand rapidly in case of mobilisation. The General Officer Commanding 3rd Division at the time, Major-General E. A. Drake-Brockman, specifically instructed his inspecting officers to report on the expansion capability of each unit serving as a nucleus, in case the government ordered a mobilisation. His request derived from a Military Board Instruction of 1936 and was, therefore, an issue of some concern to the entire force. The comments of the inspecting officers show that the 3rd Division had achieved mixed results, despite its focus on the nuclear concept since the manpower reduction of 1922.[36]

It is not possible to ascertain fully why, but the entire 15th Brigade received a negative review. For example, an inspecting officer criticised its 59th Battalion's readiness for expansion. In part this was because of its low numbers, but it was also owing to the relatively short service of its leaders and the inexperience of company commanders. As a result, he concluded, the 'standard on mobilization would not be satisfactory, and expansion and provision of leaders could not be achieved'.[37] The review of the 57th/60th Battalion was better but still somewhat negative. Its inspector noted that the battalion had problems with its leadership, including too many junior leaders who were either recent arrivals to the unit or newly promoted. In his report, the inspector wrote that the unit's 'expansion could not be achieved satisfactorily without the addition of many suitable leaders on mobilization'.[38] Similarly, the 58th Battalion failed to achieve the desired standard and was also deemed to require a large intake of trained leaders before expansion was possible.[39]

The 15th Brigade was not alone in having these problems. The Brigade Major of 4th Brigade came to similar conclusions after the inspection of his battalions. For example, observing the 14th Battalion, he noted that the unit's leaders would require a period of intensive training before they would be fit for post mobilisation duties. The 29th/22nd Battalion received an almost identical report.[40] The unfitness of these units for post mobilisation duty was a serious problem. The emphasis on leadership training under the nuclear scheme was not only to assure that the units themselves had enough leaders for their own expansion requirements, but to provide an experienced cadre for the raising of new units.

The 3rd Division's other arms showed similar shortfalls. While the inspector of the Division's medical units pointed out that the field ambulance companies did possess a number of well trained non-commissioned officers, his overall assessment concluded that these units had not reached a sufficiently high standard to be classified as efficient as a nucleus for expansion. While the non-commissioned officers of the field companies achieved a high standard, most were also nearing the end of their service careers and might be too old for field duty. Encouragingly, however, the report also noted, they could still serve as skilled instructors in training establishments.[41]

Major Les Barker's highly negative report on the 3rd Division's AASC organisation resulted in an exchange of correspondence, as the unit's commander objected to his conclusions. Barker had written, 'It is not considered that the unit has reached a sufficient standard to function as a nucleus on mobilization'.[42] The unit's commander wrote that he was surprised by Barker's report, as most of his officers had great experience and knowledge of transport and supply matters from their employment. He also insisted that they had been most diligent in the training of their men. Unconvinced, Barker wrote back that practice in peace was not the same as training for war and his main concern was that too many AASC staff did not seem to appreciate the difference. A point he highlighted was the fact, that when supporting 3rd Division camps, the AASC always operated from fixed bases with deliveries not more than half a mile from their lines.[43]

Some infantry units did manage to meet expectations. For example, the 24th Battalion received high praise. The inspector noted that, 'with few exceptions the whole of the personnel in camp has reached a sufficiently high standard to carry out their mobilization duties'.[44] However, he tempered his observation with the remark that the battalion had only 135 all ranks in camp. Of the Division's arms, only the 2nd Field Artillery Brigade received unqualified praise. Barker was favourably impressed by the entire formation and wrote, 'I consider that this Brigade has reached a standard which would enable it to function as a nucleus for expansion and post mobilization duties'.[45]

The formation's concern about the effectiveness of its artillery units reached the point that the 3rd Division participated in the trial of an experimental battery structure, the A-B-C organisation. Les Barker, the Division's Artillery Brigade Major, first proposed the scheme in 1935. Blamey supported the idea and wrote to the Military Board seeking its approval. The 3rd Division implemented the concept for the 1937 training year.[46]

The goal of the A-B-C scheme was to achieve maximum benefit from the limited funds and personnel available to the gunners, and, in particular, to increase the number of specialists in each battery to a number sufficient to support expansion. Barker pointed out that the average number of specialists in Melbourne-based batteries was only eight, and this despite the £2 annual incentive bonus.[47] His proposal called for the reorganisation of the batteries in each artillery brigade around a specific training task. Barker identified them as:

- A Battery: fully trained battery
- B Battery: specialist training battery
- C Battery: recruit training battery

Batteries retained their numerical designations but the field brigades reorganised them according to training priorities.[48]

A Battery contained the brigade's fully trained troops, and was its only operational unit. B Battery's focus was on specialist training, and consisted of those men who had completed their recruit course but needed further instruction in gun laying, range taking, signalling, or other skills. C Battery concentrated on recruit training tasks such as rifle and gun drill. C Battery was also responsible for the brigade's drivers. B and C Batteries fulfilled their training program through home parades, while A Battery was the only one which held a camp. Once recruits finished their basic instruction, the brigade assigned them to either A or B Batteries as required, or to C Battery if they were to serve as drivers.[49] Table 7 outlines the designation of 3rd Division batteries under this scheme.

Table 7[50]
A, B, C Batteries in the 3rd Division

FORMATION	A BATTERY	B BATTERY	C BATTERY
2nd Fd Bde	4th Battery	102nd Battery	5th Battery
4th Fd Bde	104th Battery	11th Battery	10th Battery
8th Fd Bde	30th Battery	108th Battery	29th Battery

The Military Board continued the experiment for the 1938 training year, to evaluate it further. However, the next year, the Military Board deemed it a failure and ordered batteries to return to traditional lines. The critical problem which doomed the experiment was the lack of numbers. To provide a full complement for A Battery, the brigades had to raid B and C Batteries for the necessary soldiers. This rendered recruit and specialist training impossible, while also failing to provide a combat-ready A Battery. For example, during 4th Field Brigade's 1938 camp, 40 per cent of A Battery's complement came from soldiers who actually belonged to B and C Batteries. The 2nd Field Brigade also had to bring A Battery up to strength with drafts from the other batteries.[51]

While numbers were the most important factor in the trial's failure, the commander of 4th Field Brigade also noted a disturbing trend in the formation's morale. Those who served in A Battery developed an air of superiority, while the men of B and C Batteries had begun to consider themselves inferior, a potential split within the formation which further hindered its development as an efficient unit. A similar problem emerged in 8th Field Brigade. The members of B Battery demonstrated a great aversion to attending home training parades, fearing that they would receive a greater share of fatigue duties. They also thought that those who did not belong to A Battery had less access to live shoots.[52]

Training

For much of the inter-war period, government attitudes and service conditions were so poor as to make meaningful training difficult. The very existence of the nucleus concept underscored the fact that the primary purpose of training was not to enhance a unit's combat effectiveness but rather to develop leaders. While large gaps exist in the surviving documentation, it is clear that throughout the inter-war period, the commanding officers of the 3rd Division faced daunting challenges in providing interesting and effective training for their units.

The rebuilt 3rd Division commenced its training program in 1921 with great promise for the future. During October and November of that year, its various arms conducted a series of voluntary evening and morning classes for their junior officers, young non-commissioned officers and selected other ranks seeking promotion. A remarkable total of 350 took part. A charitable report on the exercise commented on the absence of former AIF officers and described those who attended as 'almost wholly trainees'. Deficiencies, it admitted, were widespread, particularly amongst the infantry and engineers, and the report concluded that the 'military knowledge of the candidates was, in almost all cases, very meagre'. However, this was to be expected in the formation's first training effort, and the report went on to note some positive signs, including the great enthusiasm and desire to learn about their new trade of all who attended.[53]

However, before Gellibrand could harness the Division's enthusiasm, Hughes initiated massive changes in defence policy, and the Citizen Forces suddenly lost most of its men, budget and impetus. Shortages of men and money, and the inability to provide soldiers with realistic and rewarding training programs became the hallmark of the era. Moreover, in 1929, after Scullin's suspension of Universal Training, the army had to switch from conscription to voluntary enlistment. The result was that training became even more difficult, as officers had to dedicate greater efforts towards attracting new members and controlling expenditure. Even at the end of the inter-war period units were capable of little more than the most basic tactical manoeuvres, and the Division was unable to support training above the unit or brigade level. In fact, at no point during those 21 years of peace did the 3rd Division in its entirety take the field for a divisional exercise.

Financial limits figured prominently in the daily operation of units, as commanding officers had to find ways to stretch resources while accounting for every penny spent. As late as June 1939, the commanding officer of the 58th Battalion complained that he had to curtail training owing to the shortage of funds. This was not unusual, and the lack of money was one of the most common complaints made by unit commanders. For example, in 1932, the Division's senior officers stressed the need for more money to conduct weekend bivouacs, courses and other forms of training.[54]

12: 38th Battalion in camp, 1936. (*rear from left*) Sergeant J. Darley, Private J. McNally, unknown, Corporal J. Eddy, Private E. Lewis, unknown, Private N. Symons, Private J. McFadzean, unknown, Private J. Denyer, Private G. Waugh, unknown, unknown, unknown, Lieutenant J. Purdue, unknown (*front from left*) Private B. McNally, unknown, Corporal K. Lyons, Corporal B. Riddell, unknown, Private S. Chivers, unknown, Private P. Rosenow, Private R. Brown, unknown, Private L. Riddell. [*Courtesy* L. Riddell]

The 3rd Division Headquarters constantly reminded unit commanders of the need to economise. To that end, in 1936, the Brigade Major for the 10th Infantry Brigade, J. H. Crombie, wrote to the 24th, 37th/52nd and 39th Battalions, ordering them to survey and report on the cartage fees charged by local haulers. He was concerned that units were not securing the best possible rates. Crombie's inquiry produced a flurry of correspondence as battalion adjutants contacted local suppliers, demanded their rates, tabulated the replies and submitted reports. In another case, the Division issued detailed instructions as to who should bear the replacement cost of worn-out boots, including how to determine the reason for the equipment's failure and apportion costs if any of the damage was due to the soldier's misuse or lack of care. The circular then outlined the instances where repairs should take place or where replacement was justified.

The money available to the Division was so limited that some units exhausted their annual allotment only a few months into the training year. When that happened, it brought forth a stern rebuke from the Division headquarters. While all peacetime armies must carefully account for expenditure, the situation in Australia was so extreme that it impaired a unit's ability to conduct meaningful training.[55]

Citizen soldier training consisted of home parades at the local drill hall and an annual camp of continuous instruction. In addition, select officers and other ranks could attend special promotion courses which the 3rd Division periodically offered. A soldier's training obligation did vary during the inter-war period, and the size of the annual vote was the largest factor in determining the year's program. However, no matter how much money the government provided, the army was at no point able to train every intake under obligation. During the first half of the period, the years in which Universal Training existed, the force lacked the funds to train more than the two youngest intakes, and older soldiers attended at best a few parades.

The most important period of training was the camp of continuous instruction. From 1920 to 1929, the camp training requirement consisted of either six or eight days, except in 1921/22, when the Military Board cancelled the entire program. After 1929, the pattern continued, except that service in the army was now voluntary. Home training also varied but in most years was just four days.[56]

Despite the relatively low training obligation under the Universal Training Scheme, units had to make sure that all soldiers under obligation attended their camps. Absenteeism from camps and mandatory parades was considerable. In 1929, a summary of the 3rd Division's strength listed more soldiers on the non-effective list than appeared on its annual training establishment.[57] Commanding officers had to monitor the participation of their men and seek ways to bring defaulters into compliance. In 1928, the 3rd Division issued a directive on evasion, requiring all units to identify evaders and bring them into compliance. To assist in this quest, the division organised a special detention camp for those who failed to attend the required camp with their unit. Lieutenant-Colonel P. C. Raper, the Division's AA & QMG, instructed Area Training Officers to visit all local businesses to obtain lists of employees whose ages corresponded to the quotas subject to training. They were then to bring into training any evaders these lists revealed.[58]

With the suspension of the Universal Training Scheme, the 3rd Division lost its guaranteed, if sometimes unenthusiastic, source of soldiers. The introduction of voluntary terms of service had an immediate impact on the 3rd Division, as it became significantly more difficult for units to attract recruits and compel those in service to attend camps. A 1936 summary of attendance by other ranks at 3rd Division camps revealed that the participation rate varied between 53 and 74 per cent. The 3rd Division's overall attendance was just 61 per cent. Table 8 outlines the situation in 1936.

These figures were not an aberration. Two years later, a more limited survey of the Division's artillery brigades revealed similar attendance rates. Moreover, a five-year summary covering training years 1933/34 to 1937/38 showed that the 3rd Division's camp attendance consistently fell within the range of 61–70 per cent of posted strength.[59]

The survey outlined in Table 8 also identified whether those who failed to attend camp had been excused or whether they were absent without leave. While most personnel did seek permission to miss their camp instruction, some units suffered from strikingly high absent without leave rates. For example, nearly a third of the 46th Battalion failed to seek their commander's permission for non-attendance. In the 10th Field Company, the absent without leave rate was 30 per cent, and in the 4th Field Brigade it was 25 per cent.

Table 8
Participation by Other Ranks in 3rd Division Camps – 1936[60]

FORMATION	POSTED STRENGTH	PRESENT IN CAMP	PERCENTAGE OF STRENGTH IN CAMP
2nd Fd Bde	279	164	59%
4th Fd Bde	279	150	53%
8th Fd Bde	292	268	57%
3rd Div Eng	322	207	64%
3rd Div Sigs	209	125	59%
4th Inf Bde	1037	638	61%
10th Inf Bde	1009	636	63%
15th Inf Bde	761	463	61%
AASC	206	128	62%
AAMC	138	97	70%
AAVC	19	14	74%
Summary	4553	2792	61%

While an exact comparison of attendance rates between the Universal Training and voluntary enlistment schemes is not possible, some anecdotal evidence is available. In a report on the Division's 1923 camps of continuous instruction, it is evident that units in the past had enjoyed much higher rates of attendance. For some units, attendance actually approached 100 per cent. For example, 3rd Divisional Signals had only two absent, while only one member of the 8th Field Brigade failed to attend. The attendance rates achieved by the 10th Infantry Brigade were more typical.

10th Brigade Headquarters 80 per cent
24th Battalion 88 per cent
37th Battalion 78 per cent
39th Battalion 95 per cent
52nd Battalion 87 per cent.

While the 10th Brigade's participation rate was not as high as that achieved by the signallers, it is obvious that during Universal Training the 3rd Division was able to achieve more favourable participation rates at its camps of continuous instruction than was possible under voluntary enlistment.[61]

Commanding officers most frequently cited economic conditions as the reason why soldiers failed to participate in mandatory training activities. The Division's AASC commander noted that nearly all applications for leave cited difficulties with employment. He went so far as to contact many employers, asking for the worker's release: in a few cases, this was not done because some soldiers feared that any form of interference might jeopardise their employment.[62] The commanding officer of the 2nd Field Brigade remarked that attendance at the 1938 camp had declined to 59 per cent of posted strength, owing to the reluctance of employers to give soldiers leave.[63] Lieutenant-Colonel J. J. McKenna, the commanding officer of the 57th/60th Battalion, believed prospective recruits were afraid to join the army out of fear of having to ask their employers for leave. Lieutenant-Colonel J. E. Stevens of the Divisional Signals noted that employers' reluctance made the attainment of leave difficult.[64] The inability of the 3rd Division to bring all its members into camp had an immediate effect on its goal of training leaders and improving the operational effectiveness of the formation. This led one inspector to conclude the lack of numbers militated against efficient training.[65]

After the adoption of voluntary enlistment it emerged that, in addition to low numbers, another difficulty unit commanders faced was a high rate of turnover amongst its other ranks. This impeded the development of skilled soldiers. For example, in 1937, the average camp attendance by the 3rd Division's three field artillery brigades was 160 men. Of those, 66 had less than a year's service. This meant that relatively junior soldiers had to serve in positions which required greater experience then they possessed. The inspector noted that, in one battery, ten out of twelve drivers had never driven to a previous camp.[66]

A survey of effectiveness conducted in 1932 illustrated how great the turnover challenge was in the Division. In the years 1930–32 every arm in the formation suffered turnover rates ranging from a low of 136 per cent to a high of 218 per cent of establishment. Table 9 outlines the 3rd Division's rate of turnover.

Table 9
Turnover Within the 3rd Division, 1930-1932[67]

ARM	ESTABLISHMENT	DISCHARGES	EFFECTIVES	TURNOVER	PERCENTAGE
Artillery	875	535	796	1331	167%
Engineers	248	84	236	320	136%
Signals	198	90	191	281	147%
Infantry	2843	1539	2490	4029	162%
AASC	220	169	218	387	178%
AAMC	96	63	88	151	172%
Overall	4472	2480	4010	6490	162%

A few units suffered extremely high turnover rates. For example, the 29th/22nd Battalion replaced itself more than twice during this short period. The figures underscore the difficulty faced by the 3rd Division in developing and retaining experienced personnel. The formation was in a constant state of flux, endlessly recreating itself, as barely trained soldiers took their discharge only to be replaced by raw recruits. Nor was the 3rd Division alone in this struggle, as all Militia formations suffered from excessive turnover.

In addition to camps, the 3rd Division provided an annual series of specialist courses for soldiers interested in improving their skills. The 1937 program, for example, included advanced subjects such as signals, intelligence, survey, equitation, driving and cooking. Participation in these schools was voluntary, and was sometimes a result of an individual's desire for promotion. It also gave senior officers an opportunity to assess the leadership qualities and career prospects of junior officers and non-commissioned officers. These schools represented an important opportunity for citizen soldiers to acquire new capabilities, especially those who had decided to make a commitment to the army and saw military service as a second career.[68]

As participation in these courses was an extra duty, students had to fit them into their own time and the 3rd Division generally scheduled them on weekends or at night, obviating the need for any additional favours from employers. For example, the Division held a successful three-day course for senior officers on 27–29 January 1923, the Australia Day long weekend. In other cases, the 3rd Division took advantage of the four-day Easter Weekend.[69]

The results of these courses were mixed. The conclusions found in the report on the 1932 senior officers' course were extremely positive, noting that the officers were interested and anxious to learn, and had accomplished a great deal. The instructor remarked that most officers wanted similar courses held in the future. By contrast, the instructing staff judged a Vickers machine gun course of the same year to be merely satisfactory, admitting however that a two-day instruction only allowed a minimum of training.[70]

More worryingly, a major artillery and infantry course conducted by the 3rd Division after the abandonment of compulsory service suggested that the quality of junior students had started to decline. The course's instructors commented on the students' lack of elementary knowledge, and recommended that the 3rd Division require every officer to take an officer recruit course. A post-course assessment concluded that the standard of skill was very poor in basic subjects such as drill, small arms training and minor tactics. The artillery officers came in for particularly harsh criticism, and the report highlighted their lack of knowledge of horse management, driving and equitation, and noted that their gun drill and live shell fire were faulty. The report did compliment the non-commissioned officers for their keenness and satisfactory knowledge commensurate with their rank, but identified weaknesses in powers of command and instruction.[71]

Captain G. E. W. Hurley, the instructor for No. 1 Platoon, submitted a separate and even more damning assessment of his students. He wrote that the officers' knowledge did not approach the standard expected of persons holding commissions, and concluded that the students had made little effort to prepare for the course. He complained that they demonstrated an almost complete lack of understanding of minor tactics, that their ability to command squads was weak and their drill was poor. More significantly, Hurley stated that the students did not understand their responsibilities as officers. Another observer,

Colonel C. M. M. Dare, Commanding Officer 4th Infantry Brigade, remarked that he had been aware for some time of the declining quality of junior officers. He feared that the army was unable to attract the best sort of men and, as a consequence, officers and men of poor quality had received promotion.[72]

Throughout the army, concern over the lowering of standards reached such an extent that, in 1932, the Military Board instructed each formation to conduct a conference to obtain the views of commanding officers on the effectiveness of voluntary, in lieu of universal, training. Blamey convened the 3rd Division's conference on 30 May 1932. More than twenty officers attended, representing the entire range of the Division's corps.[73] In general, the 3rd Division's officers spoke in favour of the old system and complained that voluntary recruitment had led to a decline in unit capabilities and an increase in the difficulty of training. The officers claimed that, in the Militia, discharges were too frequent and that they maintained the strength of their units, even at reduced peacetime establishments, only with the greatest difficulty.

Lieutenant-Colonel Stanley Savige, the commander of the 24th Battalion and a future General Officer Commander of the 3rd Division, spoke harshly of the new system. He noted that the strength of his unit had declined from 480 under universal training to only 280 volunteers. More importantly, he observed, the force was failing to build a reserve, and systematic and progressive training was difficult. Although he considered his unit among Australia's finest, from a service viewpoint, he believed the 'results [were] practically worthless'. Lieutenant-Colonel R. G. Moss of the 37th/52nd Battalion believed that the physical standard of the army had declined under the Militia system and that those types who would make good officers had failed to enlist. The 4th Brigade's Colonel C. M. M. Dare, suggested that the voluntary system caused fluctuations in strength, reduced the number of soldiers passing into the reserve, promoted low unit strengths and contributed to irregular attendance. Lieutenant-Colonel H. G. Smith of the 46th Battalion highlighted the rapid turnover in personnel and concluded that low strength was due to low pay. He also observed that the administrative demands of the voluntary system reduced the time available for officer training.[74]

The officers knew that they would have to work with the voluntary system and suggested a number of ways to improve the scheme. Savige requested a new and smarter uniform, as well as the payment of travel fares. Lieutenant-Colonel Simpson, the 39th Battalion's commanding officer, thought it was vital to get the co-operation of employers. The 29th/22nd Battalion commander, Lieutenant-Colonel Frank Lind, thought that the doubling of the basic rate of pay to eight shillings a day was essential. He also called for more funds for training and repeated the demand for a better uniform.

Despite comments such as those of the commanding officer of the 58th Battalion, Lieutenant-Colonel J. Edgar, that training under the voluntary system was farcical, and numerous suggestions by other officers that compulsory training should be reintroduced, no changes took place.[75] It was not until 1936 that the government agreed to improvements in the voluntary scheme's service conditions. Until then, units had to continue to struggle within the limits of the voluntary enlistment scheme, until the onset of war forced the government to reintroduce compulsory service.

While commanding officers found voluntary enlistment less efficient than Universal Training, another factor in their dissent was that the need to find recruits added an extra burden to their already great responsibilities. In addition, officers were in the difficult position of having to track down defaulters and coerce them into fulfilling their service obligation. For many commanders, it was far simpler to discharge soldiers who had lost interest in the army. The problem reached such an extent that in 1936 the Military Board intervened. A memorandum issued by Lieutenant-Colonel N. G. Brown, the division's AA & QMG, warned unit commanders that they would be held accountable for the falling away of unit strength through excessive discharges. He informed commanding officers that they would have to provide headquarters with an explanation for any early discharges they granted. Brown considered the matter so serious that he warned that, if the reasons were not satisfactory, the 3rd Division would refer the matter to Army Headquarters, accompanied by a recommendation for the officer's discharge. Commanding officers, therefore, were in a difficult position. The government refused to provide the resources that might make participation in the army interesting and rewarding, yet still held commanders liable for failure to retain their men.[76]

In 1936, the government finally recognised that it must offer some improvements in militia service if the army were to have any chance of meeting its establishment targets. Shortly afterwards, the Military Board issued an instruction outlining the government's proposed changes in service conditions. These included a doubling of pay to eight shillings a day, the provision of fares to attend parades, and an allowance for units to hold social and athletic activities. In addition, the Military Board pledged to make funds available for units to hold weekend bivouacs as a supplementary training activity. The army also promised to renovate drill halls, amongst other commitments.[77]

However, the implementation of this important nationwide initiative was at local level, and the ensuing recruitment drive illustrated how demanding the search for new members had become on a unit's resources. The experience of the 3rd Division's 39th Battalion was typical. Its training catchment included the cities of Hawthorn and Kew. The battalion's commander approached the mayors of both communities and received their agreement to help. They set up a central co-ordinating committee and both cities sponsored public meetings to advance the battalion's cause. Notices appeared in local newspapers and unit members installed posters on trams, post offices and at strategic points. The battalion's members also distributed invitation cards to open days at its drill hall and the Kew Council provided advertising panels at tram stations.[78]

Each part of the 3rd Division made efforts to attract more soldiers. Lieutenant-Colonel Jack Stevens of the 57th/60th Battalion remembers giving numerous speeches in factories, lunch rooms and town halls.[79] 3rd Division Headquarters even required each unit to appoint a recruiting officer, whose responsibilities included arranging publicity, organising welfare and social activities such as dances, conducting meetings with employers, visiting factory sites to address workers, and liaising with civic organisations and government bodies. Blamey also appointed a divisional publicity officer, Captain L. H. Lemaire, to advance the idea of Militia service by co-ordinating promotional activities. A particularly innovative idea was arranging for each arm to have a six-minute broadcast on a local radio station. The 3rd Division also sought to improve its image by participating in highly public events, such as staffing an information booth at the Royal Agricultural Showgrounds.[80]

Despite difficulties, for most of the 3rd Division the drive produced good results. In December 1935, the Division's strength had declined to 3523 all ranks, although its establishment goal was 4416. The formation had almost a full complement of officers, but its other ranks were approximately 1000 below required levels, an important consideration, as the division needed men for the leaders to lead. In fact, the 3rd Division's strength was now just 80 per cent of establishment, one of the lowest rates of any formation in the army. However, after the recruitment drive, a strength return for December 1937 showed that the 3rd Division's numbers had risen to 4551, an improvement of 1028, nearly all of whom were other ranks, which brought the division up to 96 per cent of establishment. Oddly, despite their efforts, the 39th Battalion, did not fare as well. When the recruitment campaign began, its strength stood at just 163. Twelve months later, it had increased to nearly 350 but almost all the gains were a result of its being linked with the 37th Battalion to form the 37th/39th Battalion. The unit's membership from its centres in Kew and Hawthorn had improved, but the increase in strength was only by thirteen men.[81]

Ceremonial Activities

Although representing only a relatively small part of a citizen soldier's tasks, ceremonial activities played an important role in a unit's life. While providing some training opportunities, such as drill, ceremonial duties were principally a way to raise the profile of military service and increase a unit's visibility in its community. They also had the benefit of attracting recruits, and helped to sustain a citizen soldier's connection with his local area. Functions occurred in a public forum, typically in front of monuments or other points of central focus, and those watching could see that the men in uniform were those who, in other circumstances, played different roles in the life of the community.

Ceremonial activities took many forms. For example, in August 1939, the 58th Battalion, under the command of Lieutenant-Colonel W. G. Cannon, trooped its colours at Coburg Football Ground. At this point, the battalion operated a headquarters wing and four companies but the companies had only two platoons each. Thus, a factor in the unit's presence on the pitch was the event's recruitment potential. On one occasion, the band of the 46th Battalion, at the request of the Brighton Council, played at the opening of the community's new bath house. At another, the 3rd Division's artillery brigades had responsibility for firing the salutes on Coronation Day. Subsequently, the entire formation marched in the King's Birthday Parade. After Monash's death in 1931, the 3rd Division provided honour guards for the public memorial and units took turns accompanying the *cortège* from Parliament House to the burial site in Brighton.[82]

Perhaps the most important ongoing ceremonial activity undertaken by the 3rd Division was its participation in Anzac Day commemorations. Most units provided details for the ceremonies their local communities held. For example, in 1931, Lieutenant-Colonel Stewart took the salute of the 57th/60th Battalion's colour party at Preston Park. However, the inter-war period's most prestigious commemoration of World War I took place on Armistice Day 1934. On that day, at the 11th hour, His Royal Highness the Duke of Gloucester dedicated Melbourne's Shrine of Remembrance. The 3rd Division's units provided colour parties while their bands played for a crowd estimated at 300,000.[83]

To celebrate the State of Victoria's Centenary in 1934, the 3rd Military District held an elaborate Tattoo and Gymkhana at Flemington Show Ground. One of the 3rd Division units taking part was the 108th Field Battery, which provided a section of six guns under the command of Lieutenant D. M. Kelly. The unit had to undertake seven rehearsals in preparation for three performances, a major undertaking for its men. The Tattoo included fifteen military bands, nine from battalions of the 3rd Division: the bands of the 14th, 24th, 29th/22nd, 37th/52nd, 39th, 46th, 57th/60th, 58th and 59th Battalions. There was a demonstration of despatch riding by signallers, a night bombing attack, and a demonstration of a modern battle consisting of troops drawn from the 15th Infantry Brigade, 3rd Division Engineers and 3rd Division AAMC. At the end, all troops taking part paraded through the stadium.[84]

Sometimes, ceremonial duties took members of the Division out of Victoria to participate in events in other State capitals. In 1936, South Australia celebrated the centenary of its settlement and the 3rd Division contributed a contingent of 128 militiamen. They assembled at Victoria Barracks on 18 December, proceeded to Adelaide by train, participated in two tattoos, and marched through the city's streets before returning to Melbourne. The records failed to identify all the participants but those from the 24th Battalion were Sergeant Leslie G. Wilshire, Corporals Hilary C. Glennon and Jack Hayward, and Private Maurice W. Beaumont.[85]

Remobilisation and the Approach of War

In 1938, as the danger to the British Empire became increasingly obvious, the Australian government began to waken from its long slumber of inter-war neglect of the military. Hostile rhetoric from Nazi Germany, the Munich Crisis in Europe, and Japanese aggression in China pushed the government of Joseph Lyons into raising the Militia's establishment first to 42,000 and then to 70,000. War in Europe was still more than a year off, but the Australian government could no longer ignore the situation. By June 1939, the army numbered 77,000, a force larger than anything Australia had fielded since 1920.

In addition to the Militia's expansion, the government decided to raise a small field force of Permanent Forces personnel with which to defend Darwin. A combined arms formation of just 245 officers and men, it became known as the Darwin Mobile Force. In October 1938, Army Headquarters notified the 3rd Division of its desire to recruit some of its citizen soldiers for service in Darwin. Some 32 members of the 3rd Division came forward, and the Army accepted twenty. They became the Division's first citizen soldiers to respond to the nation's call since the end of World War I. Those volunteering came from diverse backgrounds and included Corporal Hamlet Appleby of the 3rd Division Engineers, Private George Burgess of the 59th Battalion, Gunner Francis Godsil of the 4th Field Artillery Brigade, Signaller William Harrison of the 37th/39th Battalion and Sergeant Gerald Delves of the 46th Battalion.[86]

In November 1938, the Military Board revised the 3rd Division's unit establishments, based on a force of 70,000. The new unit goals, for all ranks, were as follows:

Field Artillery Brigade: 470
Field Company: 200
Divisional Signals: 300
Infantry Battalion: 500
Divisional Supply Column: 380
Divisional Ammunition Company: 170
Field Ambulance: 170
Field Hygiene Section: 15
Mobile Veterinary Section: 30.

These establishments were actually fairly modest. After all, the revised level for an infantry battalion was still less than half that of the war requirement. However, the increases were a great improvement.

More important, the new establishments represented a change of attitude by the government. On 19 December 1938, Lyons appeared at Melbourne Town Hall to address a meeting of the Victorian Branch of the Australian Defence League. His appearance coincided with the start of a new national recruitment drive. To mark the occasion, 1600 soldiers paraded through the city's streets, 800 provided by the 3rd Division. Much remained to be done to prepare the Division for war, but at least the Lyons government had signalled that it believed the Militia was important.[87]

Sufficient records survive to illustrate the 3rd Division's expansion as a result of the latest recruitment efforts. Table 10 outlines the formation's progress. As the table shows, recruitment may have been uneven, but most units had no difficulty in reaching their new establishments and many took as many men as came forward. One of the results of the formation's expansion was that, by June 1939, the 3rd Division's artillery brigades had reraised their missing batteries.[88]

Battalions soon had to reorganise themselves and reopen centres as more recruits came forward. Before the expansion, the 46th Battalion had maintained a Headquarters Wing,

Table 10
Expansion of the 3rd Division – 1939[89]

ARM	AS OF 20 JANUARY	AS OF 24 FEBRUARY	AS OF 24 MARCH	AS OF 29 APRIL
Artillery	966	1092	1271	1290
Engineers	338	442	469	493
Signals	314	399	449	503
14th Bn	578	770	806	847
29th/22nd Bn	644	877	933	1036
46th Bn	356	469	508	547
24th Bn	264	296	317	379
37th/39th Bn	283	307	329	354
52nd Bn	945	974	1346	1398
57th/60th Bn	240	256	310	323
58th Bn	372	381	412	523
59th Bn	487	678	919	959
AASC	420	543	549	544
AAMC	217	255	256	332
AAVC	30	31	31	31
Div. Total	6515	7800	8935	9589

two infantry companies and a support company containing in total seven platoons concentrated in the Ripponlea, Brighton and Mornington sections of Melbourne. In November 1938, it put forth a plan to raise a C Company of four platoons and complete the existing companies by adding five more platoons. To accomplish this, the battalion planned to open new centres in Parkdale, Red Hill, Frankston and Hastings.[90]

Ironically, the army's expansion initially had a negative effect on the 3rd Division's readiness. The great influx of untrained men lowered the overall quality of its units, a point made by Savige, who was now commanding officer of the 10th Infantry Brigade.[91] Brigadier A. Jackson, who commanded the 15th Brigade, made a similar observation: 'large number of recruits and of new leaders has naturally retarded the progress of the units as a whole'.[92] Officers further down the chain of command echoed this theme. Divisional Signals observed that the unit's rapid expansion had caused a reversion to more elementary training.[93] The commander of the 59th Battalion complained that most of his unit's members had been in uniform for six months or less and that there was a shortage of trained personnel.[94] The 57th/60th Battalion's assessment noted that the arrival of so many new recruits had forced it to undertake a great deal more basic work.[95]

Complicating the situation, however, was that money for training remained inadequate for requirements. In his report, Jackson further observed that training funds remained insufficient and that the brigade was unable to conduct important specialist courses.[96] The 58th Battalion's commander, Lieutenant-Colonel W. G. Cannon, had to plead for more funds to enable the payment of travel fares for attendance at voluntary parades.[97] The 14th Battalion's report noted that, owing to a lack of funds, the unit was unable to let all those who wanted to participate attend a weekend training bivouac.[98]

Equipment, too, remained in short supply. Despite the request by the senior officers nearly twenty years earlier, the government had never moved to provide the force with adequate reserves of ammunition, equipment and stores that it needed for mobilisation. As a result, soldiers carried the weapons their predecessors had used in the Great War. Even simple items such as uniforms and rifles were in short supply. As units began to approach war strength, equipment was simply not available at required levels. For example, the 3rd Division identified serious shortages of night firing, anti-gas and technical signals equipment. There were difficulties also with facilities, especially rifle ranges for practice and accommodations – both fairly standard problems for any force in the midst of expansion. The 3rd Division's annual report also complained that inadequate funds and shortages of camp equipment had restricted the number of weekend bivouacs the formation could support.[99]

Despite these limitations, the 3rd Division had begun by mid-1939 to respond to the urgency of an approaching war. Much work remained before it regained the great effectiveness it had displayed on the Western Front but progress had already occurred. The 3rd Division was beginning to shed the imposed apathy of the inter-war period and twenty years of governmental neglect. The Division was again readying itself to defend the nation.

Endnotes

1 'Proposals of the Government for Home Defence of Australia', 12 June 1918, in *CPP*, vol. 4, 1917-19.

2 'Reorganisation of the Permanent Forces', 8 September 1920, AA(ACT), A2653/1, 1920, vol. 2.

3 'Report on the Military Defence of Australia, by a Conference of Senior Officers of the Australian Military Forces' (1920), p. 1, AWM1, item 20/7.

4 *Ibid.*, p. 6.

5 For the proposed divisional organisation and a list of non-divisional units, see 'Military Board Instruction, X/G 1', 13 October 1920, AA(VIC), B197, item 1937/1/23.

6 'Minutes of a Special Meeting of the Council of Defence', 12 April 1920, AA(ACT), A5954/1, item 762/19 ; and 'Military Board Agenda, No. 221/20', 17 August 1920, AA(ACT), A5954/1, item 1209/5.

7 'Minutes of Meeting of the Military Board', 9 June 1920, and 'Recommendations by the Council of Defence re Naval, Military and Air Defence', AA(ACT), A5954/1, item 1209/5.

8 H. G. Chauvel, *Report of the Inspector-General of the Australian Military Forces, 1921* (Melbourne: Government Printer, 1921), p. 14.

9 'Memorandum by the Board in Connection with the Estimates for 1920/21', 21 July 1920, AA(ACT), A5954/1, item 1209/5.

10 Data taken from 'Return Showing Approximate Strength of Units Under the New Scheme on 31st December 1920', AA(VIC), MP367, item 549/4/157; and 'Organization, Composition and Distribution of the Australian Army for War', 25 January 1928, AWM113, item 1/134.

11 'Meeting of the Military Board', 11 January 1917, AA(ACT), A2653/1, item 1917.

12 Sadler, *The Paladin*, p. 211.

13 'Preliminary Instructions for Reorganization', 13 October 1920, AA(VIC), MP367/1, item 559/1/476.

14 'Return Showing Approximate Strength of Units Under the New Scheme on 31st December 1920', AA(VIC), MP367, item 549/4/157.

15 H. G. Chauvel, *Report of the Inspector-General of the Australian Military Forces, 1922* (Melbourne: Government Printer, 1922), p. 6.

16 The story of the Singapore Strategy is told in John McCarthy, *Australia and Imperial Defence 1918-1939: A Study in Air and Sea Power* (St. Lucia: University of Queensland Press, 1976); and Ian Hamill, The Strategic Illusion: The Singapore Strategy and the Defence of Australia and New Zealand, 1919-1942 (Singapore: Singapore University Press, 1981).

17 Figures taken from 'Department of Defence, Estimates of Expenditure, 1922-23', in *CPP*, vol. 2, 1922.

18 Chauvel, *Report of the Inspector-General – 1922*, p. 9; and H. G. Chauvel, *Report of the Inspector-General of the Australian Military Forces, 1923* (Melbourne: Government Printer, 1923), p. 11.

19 H. G. Chauvel, *Report for the Inspector-General of the Australian Military Forces, 1930* (Melbourne: Government Printer, 1930), pp. 6-7.

20 'Mortar Platoon – Incorporation in Battalions', 11 March 1934, AA(VIC), MP385/3, item 106/4/292.

21 'AA & QMG to HQ 4th Infantry Brigade', 16 September 1926, AA(VIC), MP385, item 106/3/90.

22 'Organization of Infantry Battalions', 28 September 1934, AA(VIC), B1535, item 849/3/383.

23 'Proposed Reorganisation 1930/31', 4 April 1930, AA(VIC), B1535, item 849/3/53.

24 'Military Board Instruction', 16 July 1936, AWM49, item 144.

25 'Report on Camps of Continuous Training – 4th, 10th, 15th Field Companies', 15 March 1938, AWM62, item 27/4/749.

26 'Yallourn Training Centre', 20 December 1934, AA(VIC), MP385, item 106/3/249.

27 Data taken from 'Quarterly Strength Return – 3rd Division', 30 September 1931, AWM62, item 63/2/555A.

28 *Ibid.*

29 'Organisation of Battalions – 4th Infantry Brigade', 'Organisation of Battalions – 10th Infantry Brigade' and 'Organisation of Battalions – 15th Infantry Brigade', (1934), AA(VIC), MP385/3, item 106/3/240.

30 'Reorganisation – Infantry, 3rd Division', 26 May 1930, AA(VIC), B1535, item 849/3/82; and 'Organisation and Composition of Infantry Brigades', May 1930, AA(VIC), B1535, item 849/3/53.

31 'Quarterly Strength Return – 3rd Division' 30 September 1931, AWM62, item 62/2/555A; and 'Quarterly Strength Return – 3rd Division', 31 March 1935, AWM62, item 62/2/667.

32 Data taken from 'Quarterly Strength Return – 3rd Division', 30 September 1930, 31 March 1931, 30 June 1931, 30 September 1931, 31 December 1931, and 31 March 1932, all in AWM62, items 63/2/555A and 63/2/593.

33 'Dodds to Elliott', 16 March 1931, AA(VIC), B1535, item 849/3/165.

34 'Dodds to Jess', 11 March 1932, and 'Circular Letter', 6 April 1932, AA(VIC), B1535, item 849/3/165.

35 'Headquarters 15th Brigade, Organisation of Battalions', 28 August 1934, AA(VIC), MP385/3, item 106/3/240.

36 'Camp Reports', 15 February 1938, AWM62, item 27/4/749.

37 'Inspection Report – 59th Battalion', n.d., AWM62, item 27/4/749.

38 'Inspection Report – 57th/60th Battalion', 2 June 1938, AWM62, item 27/4/749.

39 'Inspection Report – 58th Battalion', n.d., AWM62, item 27/4/749.

40 'Inspection Report – 14th Battalion', 7 April 1938 and 'Inspection Report – 29th/22nd Battalion', 7 April 1938, AWM62, item 27/4/749.

41 'Report on Camp Training – AAMC 3rd Division', 5 April 1938, and 'Report on Camps of Continuous Training – 4th, 10th, 15th Fd Coys', 15 March 1938, AWM62, item 27/4/749.

42 'Inspection Report – 3rd Division, AASC', 3 March 1938, AWM62, item 27/4/749.

43 'Report on Camp Training – 3rd Division AASC', 14 and 21 March 1938, AWM62, item 27/4/749.

44 'Report on Camp – 24th Battalion', n.d., AWM62, item 27/4/749.

45 'Inspection Report – 2nd Field Brigade', 2 March 1938, AWM62, item 27/4/749.

46 'An Examination of the Organisation of Militia Field Artillery Units for Peace Training', 25 February 1935, and 'Blamey to Military Board', 10 July 1935, AA(VIC), MP385, item 106/3/325.

47 'An Examination of the Organisation of Militia Field Artillery Units for Peace Training', 25 February 1935, AA(VIC), MP385, item 106/3/325.

48 'Organisation of Field Brigades for Peace Training', 2 August 1937, AA(VIC), MP385, item 106/3/325.

49 *Ibid.*

50 *Ibid.*

51 'Inspection Report – 2nd Field Brigade', 2 March 1938, and 'Report on Training – 4th Field Brigade', 25 March 1938, AWM62, item 27/4/749; and 'Organisation for Peace Training of Field Artillery Brigades', 14 June 1938 and 'Director Military Training to Headquarters Third Division', 19 June 1939, AA(VIC), MP385, item 106/3/325.

52 'Report on Training – 4th Field Brigade', 25 March 1938, and 'Report on Camp Training – 8th Field Brigade', 15 March 1938, AWM62, item 27/4/749.

53 'Report on Training – 3rd Division, Quarter Ending 31st December 1921', AA(VIC), MP367/1, item 629/1/795.

54 'Report on Training', 13 June 1939, AA(VIC), MP70/5, item 146/1/1929; and 'Conference of Senior Officers – Precis of Written Reports', AWM49, item 107.

55 'Cartage Country Centres' 27 November 1936, AWM49, item 135; '3rd Division Circular Q34/1928: Issue of Boots to Trainees to Replace Worn-Out Boots', 25 June 1928, AWM49, item 34; and '3rd Division Circular Q56/1928: Fares and Freights', AWM49, item 35.

56 'Particulars of Training Required of Persons Liable to be Trained Under Part XII of the Defence Act Since Inception of Universal Training', AWM1, item 20/10.

57 'Half-Yearly Return of 3rd Division for Year Ending 1st February 1929', AWM62, item 63/2/555A.

58 '3rd Division Circular No. A38/1928 – Evasion of Registration', 5 July 1928, AWM49, item 34; and '3rd Division Circular No. A55/1928 – Evasion of Service, Circulars to Employers', 23 October 1928, AWM49, item 35.

59 'Camps of Continuous Training, Personnel Return – RAA 3rd Division', 1938, AWM62, item 27/4/749; and 'Papers relating to camp attendance', AWM1, item 20/10.

60 Data taken from 'Camps of Continuous Training, Personnel Return – Units of 3rd Division', 1936, AWM62, item 27/4/751. Participation by headquarters is not represented. This accounts for the discrepancy between the total for the arms and the divisional summary.

61 'Report on Training for Quarter Ending 3 March 1923', AA(VIC), MP367/1, item 629/20/68.

62 'Report on Camp of Continuous Training – AASC', n.d. AWM62, item 27/4/749.

63 'Report on Training – Camp – 1937/38 – 2nd Field Brigade', 7 March 1938, AWM62, item 27/4/749.

64 'Conference of Commanding Officers – Precis of Written Report', AWM49, item 107.

65 'Report on Camp Training – AAMC 3rd Division', 5 April 1938, AWM62, item 27/4/749.

66 'Organisation of Peace Training of Field Artillery Brigades', 6 July 1937, AA(VIC), MP385, item 106/3/325.

67 Data taken from 'Return of Effectives – 1932', AWM49, item 107.

68 'Programme of Divisional Courses', 29 October 1936, AWM49, item 37.

69 'Report on Training for Quarter Ending 3rd March 1923', AA(VIC), MP367/1, item 629/20/68.

70 *Ibid.*

71 '3rd Division Artillery and Infantry Course – Portsea: 3rd Division HQ Comments', 11 April 1932, AWM49, item 99.

72 *Ibid.*; and 'Report on Members of No. 1 Platoon – 3rd Division Artillery and Infantry Course', 15 March 1932, AWM49, item 99.

73 '3rd Division Circular', 18 May 1932, AWM49, item 107.

74 'Conference of Commanding Officers – Precis of Written Report', AWM49, item 107.

75 *Ibid.*

76 '3rd Division Memorandum, Progress in Strength of Citizen Forces', 13 November 1936, AWM49, item 37.
77 'Military Board Instruction', 16 July 1936, AWM49, item 144.
78 'Recruiting – 39th Battalion', 15 October 1936, AWM49, item 144; and 'Recruiting – 37th/39th Battalion, 1 March 1937, AWM49, item 145.
79 'Major General Sir Jack Stevens – Unpublished Manuscript', AWM, 3DRL3561.
80 'Publicity – Militia Forces, 3rd Division', 14 August 1936, 'Recruiting Officers', 17 July 1936, AWM49, item 144; and 'Militia Recruiting Information Bureau Royal Agricultural Society Show Ground', 17 September 1936, AWM49, item 37.
81 'Recruiting Return of Enlistments', 12 June 1937, AWM49, item 145; 'Quarterly Strength Return – 3rd Division', 31 December 1937, AWM62, item 63/2/768; and 'Quarterly Strength Return – Militia Forces, 31 December 1935', AWM62, item 63/2/699.
82 'Program – 58th Battalion', 1939, AWM62, item 47/10/133; '46th Battalion to HQ 4th Infantry Brigade', 5 December 1938, AWM49, item 130; 'Artillery Salutes', AWM62, item 47/6/82; and Robin S. Corfield, *Hold Hard Cobbers, The Story of the 57th and 60th and 57/60th Australian Infantry Battalions 1912-1990*, vol. 2, *1930-1990* (Glenhuntly, VIC.: 57/60th Battalion Association, 1991), p. 16.
83 Corfield, *Hold Hard Cobbers*, pp. 15, 21.
84 'Services' Tattoo and Gymkhana', 4 August 1934, and 'Services' Centenary Council Circular No. 2', 1 August 1934, AWM49, item 175.
85 'Centenary Celebration, South Australia: Victorian Contingent', AWM49, item 136.
86 'Applications to Enlist in Darwin Mobile Force – Successful', AWM62, item 61/2/1961.
87 'Recruitment Campaign', 30 November 1938, AWM49, item 130.
88 Data taken from 'Progress Strength Return No. 10', 24 February 1939, and 'Progress Strength Return No. 5', 20 January 1939, AWM62, item 63/2/810, part 2; and 'Progress Strength Return No. 14', 24 March 1939, and 'Progress Strength Return No. 19', 29 April 1939, AWM62, item 63/2/810, part 1.
89 'Proposed Reorganisation 46th Battalion', 30 November 1938, AWM49, item 130.
90 'Training Report, 1938/39 – 3rd Division', 23 June 1939, AA(VIC), MP70/5, item 146/1/1929.
91 'Training Report – 10th Infantry Brigade', 2 June 1939, AA(VIC), MP70/5, item 146/1/1929.
92 'Annual Training Report – 15th Infantry Brigade', 21 June 1939, AA(VIC), MP70/5, item 146/1/1929.
93 'Annual Training Report: Part I, Signals', 21 June 1939, AA(VIC), MP70/5, item 146/1/1929.
94 'Training Report – 59th Battalion', 20 June 1939, AA(VIC), MP70/5, item 146/1/1929.
95 'Annual Training Report – 57th/60th Battalion', 27 May 1939, AA(VIC), MP70/5, item 146/1/1929.
96 'Annual Training Report – 15th Infantry Brigade', 21 June 1939, AA(VIC), MP70/5, item 146/1/1929.
97 'Report on Training – 58th Battalion', 13 June 1939, AA(VIC), MP70/5, item 146/1/1929.
98 'Annual Report – 14th Battalion', 24 May 1939, AA(VIC), MP70/5, item 146/1/1929.
99 'Training Report, 1938/39 – 3rd Division', 23 June 1939, AA(VIC), MP70/5, item 146/1/1929.

Chapter 4
The 3rd Australian Division in World War II

The Return to War

In September 1938 Britain's Prime Minister, Neville Chamberlain, returned to London from Munich after meeting Adolph Hitler, the Chancellor of Nazi Germany. Arriving home, he proclaimed that he had secured 'peace with honour ... peace for our time'.[1] The worsening situation on the Continent quickly proved the fallacy of his claim.

At Munich, Chamberlain had joined France's leader Édouard Daladier in acquiescing to Germany's annexation of the Sudetenland region of Czechoslovakia, in exchange for a pledge of peace. Hitler soon revealed his future territorial ambitions and his willingness to plunge Europe into war in order to establish Germany's hegemony. Only a few months after having agreed to leave the rest of Czechoslovakia alone, Hitler seized the remainder of the country by force. Poland was his next target and, on 1 September 1939, German troops crossed the Polish-German frontier. Britain and France, having guaranteed Poland's security, demanded Germany's immediate withdrawal. Undeterred, German troops continued their onslaught. On 3 September, as the ultimatum expired, Chamberlain announced that Britain would honour its commitment to Poland and declared war on Germany. A few hours later, Australia's Prime Minister, Robert Menzies, addressed the nation and confirmed that Australia, too, was at war.

Mobilisation

Surprisingly, the declaration of war had little immediate effect on the development of the Militia. Although Menzies had not hesitated to bring Australia into the conflict, he took a measured approach to its preparations and commitments for war. Instead of declaring a full-scale mobilisation of the army, he authorised the call-out of only 8000 militiamen for full-time duty, and that for just sixteen days. Most members of the Militia continued to carry out their civilian tasks in their own communities. The only difference attributable to the start of the war was that the army issued them with a 'Mobilization Assembly Card', which identified them as part of the Militia, and, in the case of mobilisation, informed them which day to report and what equipment to bring. Attached to the card was a travel voucher for free rail passage, if required, to the reporting depot.[2]

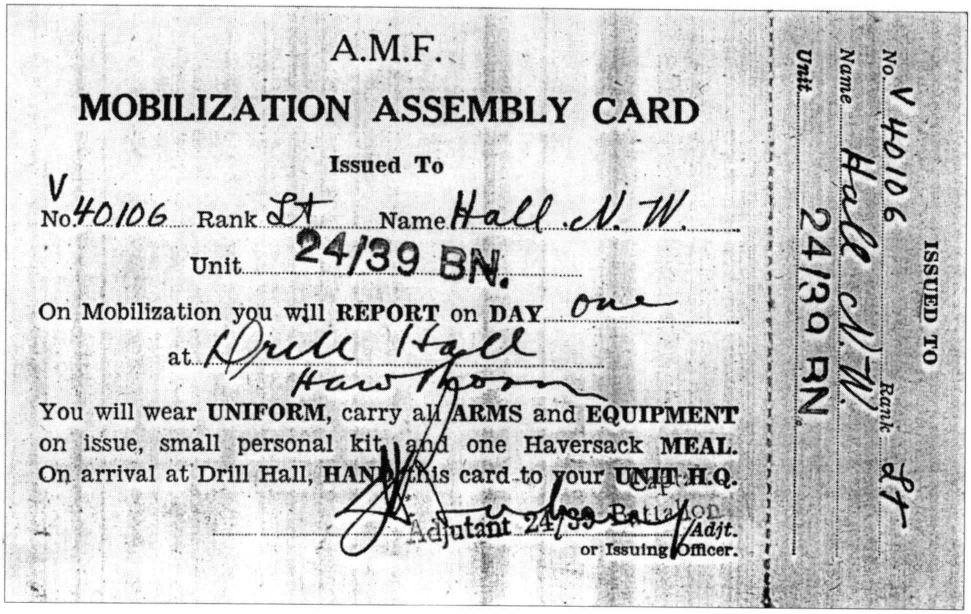

A.M.F.

MOBILIZATION ASSEMBLY CARD

Issued To

No. V 40106 Rank Lt Name Hall N. W.

Unit 24/39 BN.

On Mobilization you will **REPORT** on **DAY** one

at Drill Hall Hawthorn

You will wear **UNIFORM**, carry all **ARMS** and **EQUIPMENT** on issue, small personal kit, and one Haversack **MEAL**. On arrival at Drill Hall, **HAND** this card to your **UNIT H.Q.**

Capt.
Adjutant 24/39 Battalion
Adjt.
or Issuing Officer.

ISSUED TO

No. V 40106

Name Hall N. W. *Rank* Lt

Unit 24/39 BN

13: Lieutenant N. W. Hall's mobilization card (front).

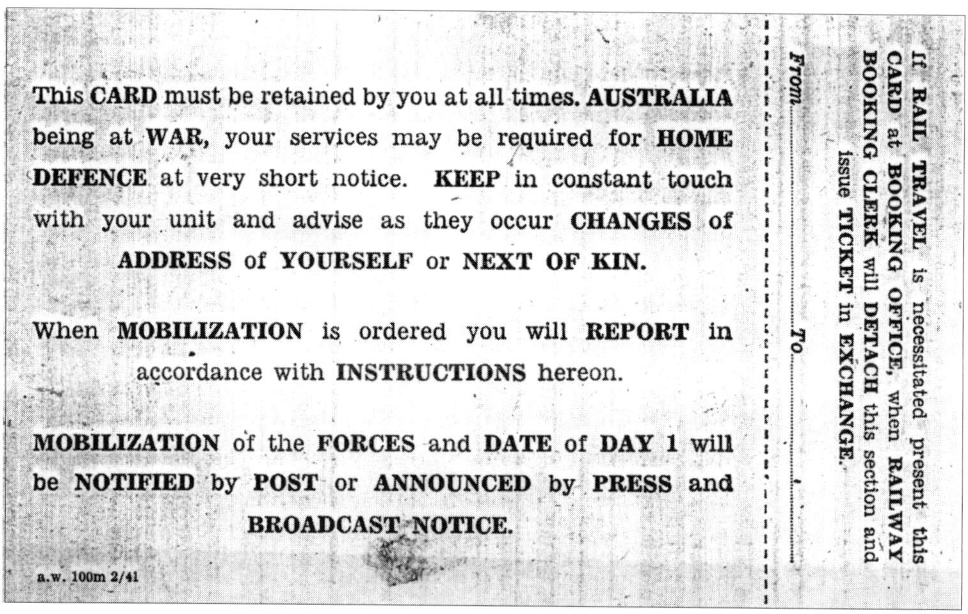

This **CARD** must be retained by you at all times. **AUSTRALIA** being at **WAR**, your services may be required for **HOME DEFENCE** at very short notice. **KEEP** in constant touch with your unit and advise as they occur **CHANGES** of **ADDRESS** of **YOURSELF** or **NEXT OF KIN**.

When **MOBILIZATION** is ordered you will **REPORT** in accordance with **INSTRUCTIONS** hereon.

MOBILIZATION of the **FORCES** and **DATE** of **DAY 1** will be **NOTIFIED** by **POST** or **ANNOUNCED** by **PRESS** and **BROADCAST NOTICE.**

a.w. 100m 2/41

If **RAIL TRAVEL** is necessitated present this **CARD** at **BOOKING OFFICE**, when **RAILWAY BOOKING CLERK** will **DETACH** this section and issue **TICKET** in **EXCHANGE**.

From .... *To* ....

14: Lieutenant N. W. Hall's mobilization card (back).

For those called to duty, their purpose was to guard vulnerable points, provide internal security, and help bring the nation's coastal and aerial defences to full strength. For example, a small party drawn from the 57th/60th Battalion, under the command of Captain F. G. Pullen, spent the first few days of the war protecting the radio masts near Mont Park. Another early participant was Lieutenant N. W. Hall of the 39th Battalion, who led a detachment charged with rounding up Germans for internment. His orders were to deliver the internees to the grounds of the Wirth's Circus, across the river from Flinders Street Railway Station. As there was no security at that location, Hall obtained a supply of barbed wire and told the Germans, to 'wire yourselves in'.[3]

One of the first, and youngest, citizen soldiers called up was Athol L. Cottrill, a sixteen-year old member of the Caulfield Technical College Cadets. The cadets were affiliated to the 37th Battalion. When Menzies declared war, Cottrill received a telegram ordering him to report to the 37th Battalion's drill hall, where he was attached to a Militia section, and spent the day helping round up male German citizens. Everything was moving quickly and no one seemed to notice his age. On his second day of service, he helped march a party of Germans to the Flinders Street Railway Station. There, the Militia put them on a train whose windows had been sealed with barbed wire. Cottrill accompanied the Germans to Tatura, where the government had opened an internment camp. There he served as a guard, and even helped recapture six Germans who broke out of camp. On being relieved by the 24th/39th Battalion, Cottrill's detachment reported to a training camp at Trawool where he joined a machine gun crew in the support company. At that time, his military career came to an end, at least temporarily. His parents had become concerned at the increasing seriousness of the situation and notified the government of his tender age, and he was summarily discharged. The next year, however, he lied about his age and joined the RAAF.[4]

By the middle of September, the 3rd Division had 700 officers and other ranks serving on security details. Their assignments varied widely. 3rd Division Signals provided detachments of men to Victoria Barracks and Port Phillip Fortress to assist with communications. They also staffed a special intercept cell responsible for gathering intelligence from radio broadcasts from South East Asia. The 24th/39th Battalion received the task of guarding the Internment Reception Camp at Wirth's Park and the permanent camp near Tatura. Headquarters 3rd District Base drew on the 3rd Division's infantry to form an ad hoc metropolitan security battalion, allocating troops to key points in Melbourne, such as the Essendon Aerodrome. The 29th Battalion posted men at the Commonwealth Oil Refinery and the Widdis Battery Factory. Other soldiers protected military installations as far away as Seymour and Albury. Gradually the army raised a number of static garrison battalions which took over these security duties, allowing the seconded 3rd Division men to return to their units.[5]

Menzies did not authorise a call-up of the entire Militia until 15 September, and then for only a month-long period of training in two batches each of 40,000 men. In such a brief period, the army could hope to accomplish little more than individual and platoon training, and possibly to introduce company and battalion skills. Before that training could commence, however, each unit held a special instructor's course for officers, non-commissioned officers and selected privates. The purpose of this course was to prepare the attendees to carry out the syllabus for the month-long training period. Upon its completion, many participants received promotion. Captain Hugh Wrigley conducted the 37th Battalion's course at the Sale Showground, and more than 90 officers and other ranks participated.[6] That there was a need for such courses demonstrated the weaknesses of the nuclear army system and the failure of most units to reach a state of effectiveness for mobilisation.[7]

15: Athol Cottrill at camp with the 37th Battalion. The driver is R. Deeker of the 24th Battalion. [*Courtesy* Athol Cottrill]

In January 1940, the government authorised a further period of Militia training, this time for three months. This became the pattern for the next two years, as citizen soldiers alternated periods of military training with civilian employment. The three-month program allowed for a more ambitious training agenda. After a month of individual and sub-unit training, the syllabus called for unit and brigade exercises. The scheme fell short of reaching divisional level exercises, although it noted the possibility. At that time, Menzies also reintroduced Universal Service, which had been suspended since 1929.[8] All unmarried men who turned 21 in the year ending 30 June 1940 had to fulfil a three-month training obligation before passing into the reserves. In March, the Militia commenced another three-month cycle.

Those who marched into camp for the first time saw proof of the government's long period of inter-war neglect. Essential equipment and supplies were scarce, and many recruits had to train with sticks instead of rifles. Most weapons were of World War I vintage, some in such decrepit condition that constant repair was necessary. In the 24th Battalion, all transport was horse drawn. Clarence W. Wood, of the 3rd Division AASC, remembers using a horse-drawn general service wagon to pick up supplies from the railway station and distribute them to unit cook houses. Even simple items such as tents were in critically short supply. In March 1940, the 3rd Division Headquarters issued a warning stating that the wastage rate of tents exceeded new production. The memorandum ordered officers to pay closer attention to the preservation of their unit's tents. The government had attempted to make up the Militia's equipment deficiencies by placing orders with factories in Britain, but the British Army had first call on those materials and nothing arrived in time for the commencement of training.[9]

The 24th Battalion's camp of August–October 1941 demonstrated that improvisation was a necessary part of a unit's training. The shortage of transport meant that the troops practised embussing and debussing from imaginary vehicles. The battalion's Padre, A. E. Albiston, a Salvation Army Chaplain, remembers becoming a 'casualty' during an exercise in which he was hit by imaginary fire from an imaginary gun. In another instance, when the battalion was practising for a dawn attack, the instructing sergeant outlined the situation by explaining that a line of trees, for the purpose of the exercise, was now the enemy; the saplings the soldiers held were rifles, and some nearby tree trunks became a battery of 25 pounders. The sergeant then asked Albiston to leave the area, as he was standing in a danger zone. The padre, obviously a quick wit, replied, 'For the purpose of the exercise, I am here to bury the dead'.[10]

Individual memories of the time at Seymour are mixed, not surprisingly given the number of soldiers in camp and the great range of their experiences. The 3rd Division contained a high percentage of eighteen and nineteen year olds, for whom the camp at Seymour was their first time away from parents and home. Signaller N. H. Taylor, described it as a strange experience. The men had to make some adjustment to army life. They slept in tents on straw palliasses. Food was ample but of uncertain quality. Gordon Huggard remembers the heat most. He described the camp as 'hot, dry and dusty', with its buildings clad in hot tin sides with hot metal roofs. For George H. Carter, the most vivid memory was of the dust which covered one's bread, butter, jam and tea with a layer of fine red powder. However, as Saul Spielman pointed out, it was also an opportunity to learn new skills. A. E. Doyle of the AASC learned how to drive trucks and repair engines, while John C. Thomas mastered the motor bike and became a dispatch rider in 3rd Division Signals.[11]

Having declared war, Menzies hesitated over whether to provide an expeditionary force for service in Europe. He could not ignore the requirements of the active theatre for long, however, and Australia soon offered Britain a force of one division. This became the 6th Australian Infantry Division, the first formation raised for what was to become known as the 2nd Australian Imperial Force (2nd AIF). As in World War I the army raised the 6th Division on a regional basis from volunteers. This meant that the government had to form its overseas force from outside the existing Militia organisation. Menzies had little choice. The Militia existed to defend Australia, and the *Defence Act* did not allow the government to send its members overseas without first obtaining each individual's explicit consent. Reinforcing the government's reluctance to deploy the Militia, however, was the uncertainty as to the threat posed by Japan. The Minister for the Army, G. A. Street, advocated giving priority to the Militia, and Menzies announced the government's intention to maintain its citizen forces at a strength of approximately 75,000. The government's official policy in the war's opening months was to discourage Militia soldiers from volunteering for the 2nd AIF.[12]

Despite the government's initial intentions, the demands of war soon proved impossible to ignore. Raising the 2nd AIF quickly became the nation's primary military obligation, and the government authorised three further divisions for overseas service – the 7th, 8th and 9th Australian Infantry Divisions – as well as a corps headquarters and corps troops. For the 3rd Division, the raising of the 2nd AIF represented a curious twist of fate. During World War I the 3rd Division had been a part of the AIF and had served with distinction on the Western Front. However, when the government re-established the Army after 1918, it had the specific intention of creating a force for national defence, not overseas service. Individual soldiers could volunteer for the 2nd AIF, but the 3rd Division, barring changes in the *Defence Act* which the government would not entertain, had to remain at home.

Australia once again had to maintain a two-army system, as it had in World War I: an expeditionary force in the Middle East and a Home Army for continental defence. The difference in World War II was that the 3rd Division was now part of the Home Army. This fact had serious consequences for the 3rd Division's development and, in particular, inhibited the efforts of its leaders to improve its effectiveness during the first part of the war.

Although the government was anxious to maintain the strength of the Militia, the 3rd Division still provided many volunteers for each of the AIF's divisions. In addition, its soldiers also joined the RAN or RAAF, as those services were also expanded. In one instance, nearly an entire battalion of the 3rd Division resolved en masse to serve with the 2nd AIF. Shortly after the declaration of war, the commanding officer of the 37th battalion, Lieutenant-Colonel A. W. Stewart, and 90 per cent of his men offered themselves as a complete unit: the offer was refused. In 1940, a majority of the men in the 4th Infantry Brigade voted to join the AIF as a formation but the Army also let that opportunity pass.[13]

Once the government relaxed its objection to citizen soldiers volunteering, the Militia became a de-facto training depot for the expeditionary force. Not only did the Militia provide a source of partially trained other ranks, but the rapidly expanding AIF desperately needed experienced men for leadership positions, and only the Militia could meet the requirement. The pattern followed by the 24th/39th Battalion underscored the situation.

In October 1939, the battalion's commander, Lieutenant-Colonel Theo Walker, received command of the 2nd/7th Battalion in the new 6th Division. He took with him about half of the 24th/39th Battalion's officers and many of its other ranks. His successor, Lieutenant-Colonel Alan Spowers, then raised the 7th Division's 2nd/24th Battalion. He was briefly followed by Major Clarrie Owen, who also soon left the 24th/39th to form the 2nd/22nd Battalion in the 8th Division. Events in the 57th/60th Battalion followed a similar course. After the war's commencement, the battalion's commander, Lieutenant-Colonel Jack Stevens, was one of the first citizen soldiers to volunteer for the 2nd AIF. His successor, Major Bernard Evans, remained only a little longer before leaving to organise the 2nd/23rd Battalion. They took with them many of the battalion's officers and non-commissioned officers. The need to fill leadership positions in the AIF saw Militia officers receive advancement in the expeditionary force. Major W. G. Cannon of the 29th Battalion became a Lieutenant-Colonel and commander of the 2nd/14th Battalion. Also leaving the 29th Battalion with Cannon was Captain Steve Noonan, who received a promotion to Major. The volunteering of 3rd Division soldiers was not limited to the infantry but was widespread throughout the formation. In 1941, Saul Spielman was a sergeant with 3rd Division Signals, and remembered that, when the army raised the 8th Infantry Division AIF, the core of its signal unit came from 3rd Division volunteers.[14]

As a consequence of the citizen soldiers' desire for active service, each time the army raised a new AIF unit, the 3rd Division lost a large part of its best trained personnel. The depletion in the Militia ranks became such a serious concern that the government temporarily prohibited the transfer of personnel between the two forces. When it lifted the restriction in 1942, militiamen once again enlisted in droves in the AIF. While the drain in manpower was itself serious, of even greater significance was the effect on the Militia's capabilities. In a situation reminiscent of the inter-war period, unit commanders again had to struggle against a high turnover rate to maintain the strength of their units. In compensation, the 3rd Division periodically received large drafts of new men, in a futile attempt to keep unit strengths at the required levels. For example, Private A. Les Jones marched into the 3rd Division's camp on 2 August 1941 with 1000 other new soldiers. In January 1942, further drafts of recently inducted men arrived to strengthen the formation. As the more experienced and older men volunteered for overseas service, the average age of units declined and, by 1942, the Militia consisted largely of eighteen-year olds barely out of school and with little broader experience. Unit training cycles were also disrupted by the periodic arrival of inexperienced national servicemen who remained with the unit for only a single training cycle before disappearing into the reserves. Thus, the loss of trained men, the lack of equipment, the constant arrival of new personnel, and the inability of units to maintain their war establishments left the 3rd Division in a continuous state of rebuilding. While progress was constantly made in individual skills, the 3rd Division was still not fit for field operations even after two years of war.[15]

If Australia's need to support two distinct armies was largely unavoidable, given the *Defence Act*, government policy exacerbated the situation and helped foster an acrimonious division between the force's different branches. While the tension that developed between the 2nd AIF and the Militia was a reflection of the traditional derision all fighting troops direct towards those who remain at home, the critics failed to appreciate the difficulties citizen soldier units faced and ignored the contribution they made to the leadership and staffing of the expeditionary force. Moreover, the split within the force denied the reality that many Militia soldiers were also AIF soldiers who continued to serve throughout the war in Militia units. In fact, most soldiers serving in the 3rd Division were also members of the AIF. For example, Spielman opted for the AIF in 1942 but remained with 3rd Division Signals for the rest of the war. Harry Finn of the 3rd Division AASC was another man in this situation. He joined the citizen forces in 1940, volunteered for the AIF in 1942 and spent 920 days outside Australia. Some citizen soldiers managed to serve in both branches of the army. R. F. C. Campbell began the war as a citizen soldier in 3rd Division Signals, volunteered for the 2nd AIF and spent most of the war with the 1st Australian Armoured Division. After its disbandment he rejoined 3rd Division on Bougainville in early 1945.[16]

The government's ban on the enlistment of militiamen into the AIF was imposed out of the necessity to maintain the strength of the citizen forces. Otherwise, even more members of the 3rd Division would have enlisted in the AIF. For example, when Prime Minister John Curtin removed the restriction in June 1942, 45 per cent of the 58th Battalion and 57 per cent of the 59th Battalion immediately joined the AIF. This did not change the unit's composition, however, as these men continued to serve with their battalions.[17]

Once 65 per cent of the members of a Militia unit had volunteered for the AIF, the army added the term AIF to its designation. For example, when the 3rd Division's signal unit passed the threshold, its correct title was '3rd Australian Division Signals, AIF'. The surviving records are unclear as to how many of the 3rd Division's units bore an AIF identification, but it is clear that most did. Unfortunately, the change of title had little effect on the status of these units in the eyes of higher authorities. From the perspective of the government's operational policy the 6th–9th Divisions were the AIF, while the 3rd Division remained a Militia formation even though most of its members were also AIF soldiers.

The AIF employed several terms of derision to demonstrate their perceived sense of superiority over the Militia. The most common was 'Chockos', a revival of a World War I insult denoting a chocolate soldier, one who carried chocolates in his pack instead of ammunition and melted in the heat of battle. Another favourite was 'koalas', a reference to an Australian law prohibiting the shooting or exporting of these marsupials. For those risking their lives, there may have been validity in these taunts, but their appropriateness declined rapidly after the entry of Japan into the war, when Militia units found themselves fighting side by side with the 2nd AIF in the jungles of New Guinea.

What is more questionable was the government's continued support for the two-army system, even after the Militia fulfilled its duty as the defender of Australia upon the war's spread to the Pacific. Moreover, the Australian government appeared to go out of its way to emphasise the differences between the two branches. The Home Army and the 2nd AIF developed as entirely different organisations, with separate chains of command, structures and even colour patches. Their terms of service also differed; the Treasury taxed Militia pay, and charged citizen soldiers for their postage, unlike their AIF counterparts. Militia units also had to have dry canteens, at least officially, unlike those belonging to the AIF.[18]

The split within the army should never have happened, nor should the government have allowed it to continue once Japan had entered the war. Once the Militia was committed to combat in New Guinea, the government's continued support of the two-army system became hypocritical. Unfortunately, Australia's wartime prime ministers lacked the courage to heal the breach. The two-army system did not increase Australia's military strength, it only tarnished the service of those soldiers who chose to defend their country by serving in the Militia.

Japan Strikes

On 7 December 1941, at 6 a.m. Hawaii time, Japanese forces bombed the American fleet at Pearl Harbor and, shortly thereafter, struck other targets in the Philippines, on the Malayan Peninsula and elsewhere. British resistance in Malaya collapsed quickly and, on 15 February 1942 Singapore surrendered. With Singapore's fall, most of Australia's 8th Infantry Division marched into captivity. The Japanese continued their offensive, pushing onwards to the islands of the Dutch East Indies, to Australia's north. A separate enemy drive saw the fall of Rabaul in New Britain on 23 January and in, early March, Japanese troops landed at Lae and Salamaua in New Guinea. In the midst of its desperate battle for survival against Nazi Germany, Britain had attempted to fulfil its pre-war security guarantee to Australia by sending a Royal Navy task force based on the battleship *Prince of Wales* and the battle-cruiser *Repulse* to Singapore. On 10 December 1941, Japanese planes located both ships and sent them to the bottom. Within a few weeks of the war's spread to the Pacific the government's pre-war defence plan had crumbled, Japanese forces had occupied Australian territory in New Guinea and the entire continent seemed under imminent threat of invasion. With the 2nd AIF committed to the Middle East, the only force available for the defence of the homeland was the Militia.

At the time of Japan's attack, the 3rd Division was in camp at Seymour. It had recently converted its organisation from the pre-war Australian standard to the Imperial standard. Its brigades now had three instead of four battalions and the artillery brigades had become regiments. The 3rd Division had also gained some specialist elements, such as a reconnaissance battalion and anti-tank batteries. This was the first of many modifications to the 3rd Division structure, producing a wartime formation considerably different from its pre-war counterpart. Table 11 outlines the 3rd Division's organisation, and the location of its units around Seymour, in January 1942.

Table 11
3rd Division Order of Battle with Training Locations, January 1942[19]

FORMATION	LOCATION	SUB-ELEMENTS
HQ 3rd Division	Rokeby	HQ RAA (incl HQ 6th A/Tk Regt, 21st A/Tk Bty; and 263rd LAD)
		HQ RAE and 53rd Fd Park Coy
		HQ 3rd Div Sigs and No. 1 Coy
		HQ AASC
		Employment Pl
		Div Sec Int Corps
		HQ Div Provost Coy
		HQ Div Postal Unit
		3rd Fd Hyg Sec
	Webb's Paddock	3rd Div Recce Bn
4th Bde Gp	Rifle Range, Seymour	HQ 4th Inf Bde (incl Sig Sec and LAD)
		22nd, 29th, 46th Bns
		4th Fd Regt, Sig Sec, and LAD
		22nd A/Tk Bty
		4th Fd Coy
		4th Fd Amb
		Employment Pl
		Det Provost Coy
		Sec Div Postal Unit
10th Bde Gp	Camp Site 2	HQ 10th Inf Bde (incl Sig Sec and LAD)
		24th, 37th, 52nd Bns
		2nd Fd Regt, Sig Sec, and LAD
		23rd A/Tk Bty
		10th Fd Coy
		10th Fd Amb
		Employment Pl
		Det Provost Coy
		Sec Div Postal Unit
15th Bde Gp	Camp Site 29A	HQ 15th Inf Bde (incl Sig Sec and LAD)
		Employment Pl
		Det Provost Coy
		Sec Div Postal Unit
	Camp Site 25A	57th/60th Bn
	Camp Site 20A	58th Bn
	Camp Site 30	59th Bn
	Camp Site 17	8th Fd Regt, Sig Sec, and LAD
		24th A/Tk Bty
		15th Fd Amb
	Camp Site 25	15th Fd Coy
AASC	Camp Site 17	3rd Div Supply Col
		3rd Div Ammunition Coy
		3rd Div Petrol Coy

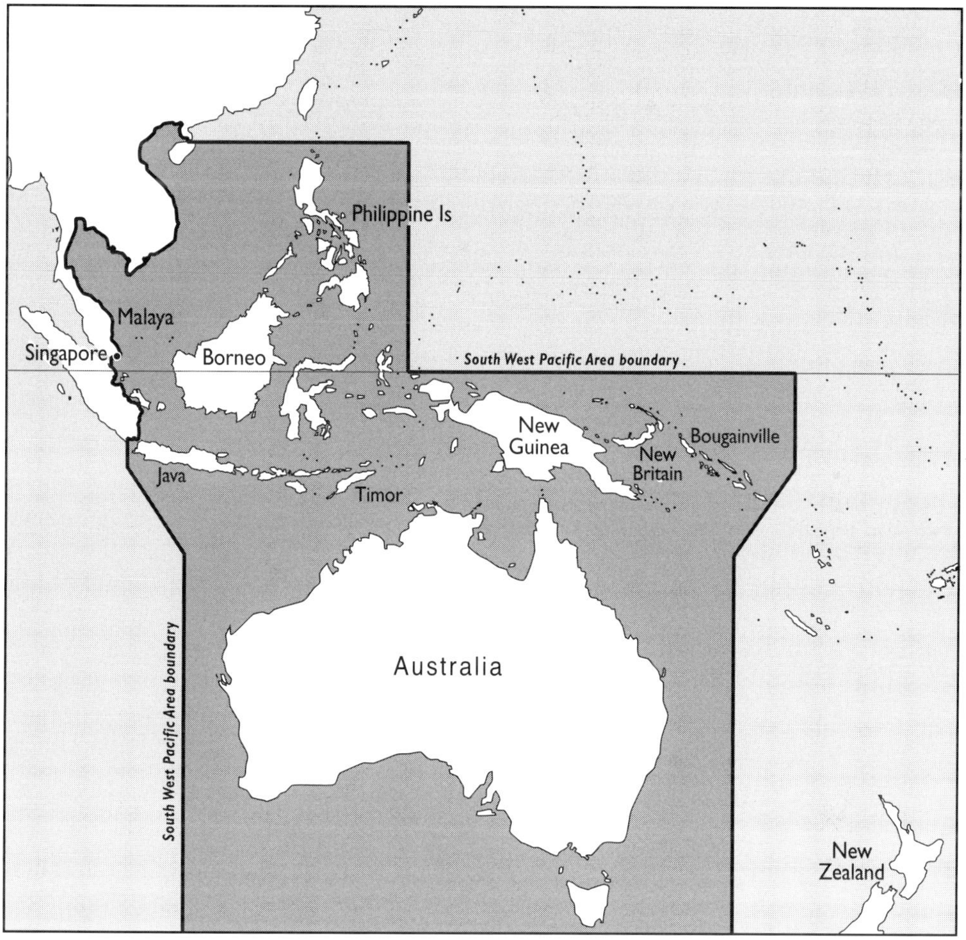

MAP 4: The South West Pacific Area

The apparent completeness of the 3rd Division's order of battle masked considerable deficiencies. The formation was actually far from ready for field operations, and its effectiveness differed little from its pre-war nuclear status. To date, training had consisted largely of individual skills instruction and route marches. The Division's actual strength was barely 50 per cent of its war establishment. It had proved impossible for units to maintain numbers, as Militia soldiers continued to depart for the AIF. Many of its members were recently conscripted youths just out of school, and few soldiers or leaders had practical experience. Lieutenant-Colonel John Wilton, soon to become the formation's GSO1, expressed to his wife the fear that the Division was young and that it had few 'old sweat' types, most of whom were in the AIF.[20]

Another grave problem impairing the quality of the 3rd Division's training efforts was its chronic shortage of modern equipment. On 8 December 1941, the Headquarters of 4th Infantry Brigade complained that 'difficulty in obtaining tng [training] stores and equipment is hindering tng [training]'.[21] So bad was the situation that, on 1 April 1942, the quartermaster of 37th Battalion, Lieutenant John Lloyd, noted the unit's augmentation of firepower by the receipt of three Vickers and six Lewis machine guns. The Lewis guns, of course, were World War I surplus, and the battalion apparently was not yet eligible for the modern Bren gun. Lieutenant Victor Cole, the 37th Battalion's transport officer, made what must have been a fairly common complaint among those serving in Militia battalions: that the unit was so far below its vehicle war establishment that the deficiency affected its training. Cole explained that lack of vehicles made movement difficult and time consuming as the limited transport that was available had to ferry troops to and from their destinations. A report prepared by Headquarters 15th Brigade made a similar observation. It complained that the formation's lack of transport was seriously handicapping its training program.[22]

Equipment problems remained intractable. Well into 1942 the mortar platoon of the 37th/52nd Battalion used an improvised weapon made from galvanised water pipe. When the 3rd Division AASC left Victoria for the move to Queensland, many of its soldiers still had broomsticks for weapons. A lucky few carried Boer War surplus rifles with just five rounds of ammunition apiece. The problem the Militia faced regarding equipment shortages was the result of two related government policies. First, it was a legacy of the government's short-sightedness during the inter-war period and its failure to stock reserves of materials adequate not just for mobilisation but also for expansion. Second, it was the harsh reality of a situation where the 2nd AIF had first call on any new production.[23]

While the 3rd Division's adverse training environment was not fully under its control, its senior officers must also accept some of the responsibility for the formation's slowness to progress into an effective combat organisation. Even at the end of 1941, training continued at a leisurely pace more suited for peacetime than for a war of national survival. For example, the 24th Battalion's war diary entries for the period 4 October 1941–4 January 1942 show that, of the 93 available days, the unit did not train on 33 and for 35.5 per cent of their time at Seymour, the men either had leave or were engaged in camp maintenance or recreational activities. Leave alone consumed 25 days (26.9 per cent). It was not until nearly a fortnight after Japan had unleashed its assault that the 24th Battalion moved to what its war diary describes as an intensive training period. For the first time, the battalion experienced a seven-day, 24-hour, training week. Despite the change in policy, however, the diary's entries for 20 and 21 December still recorded the granting of local leave each day to 25 per cent of the battalion.[24]

The other ranks also failed to appreciate the seriousness of the situation. As Christmas Day 1941 approached, they objected to the intensification of the training regime. The issue reached the point that a delegation of men, drawn from across the entire 10th Infantry Brigade, made a formal petition for holiday leave, going as far as attempting to approach brigade headquarters with their request. The 24th Battalion's war diary records the denial of their petition and a request that the men withdraw their demand. The exact resolution of the matter is not known. The men did have Christmas Day largely to themselves, and the brigade headquarters allowed visitors into camp. However, the men spent Boxing Day at the firing range.

On 9 January 1942, Major-General Stanley Savige took over from Drake-Brockman as the 3rd Division's commander. He was a good choice, having recently returned to Australia from the Middle East after serving as a brigade commander with the AIF in the Libyan, Greek and Syrian campaigns. He had also been a citizen soldier during the inter-war period and had commanded the 3rd Division's 24th and 37th Battalions, as well as the 10th Infantry Brigade. He was also a World War I veteran. It was now up to this highly experienced officer to get the 3rd Division ready for war.

Savige was not pleased by what he found at Seymour. It was not just the fact that the 3rd Division was under strength and inexperienced that disturbed Savige. Rather, it was the quality of the formation's officers and the atmosphere of peacetime lassitude pervading training efforts. He believed that the time was long overdue for the 3rd Division to replace its lackadaisical approach to training with the seriousness of purpose that was essential if the formation were to succeed in battle.[25]

In a letter of 17 July 1943 to one of his brigade commanders, Brigadier H. H. Hammer, Savige wrote frankly of the attitude which greeted him when he had first joined the Division more than a year earlier. He commented, 'When I got back I was really sick at heart when I saw the unreal outlook & effort which I could only term as "Gathering mushrooms & chasing rabbits"'. He went on, 'Bullshit, malingering, social ambition and bugger all in the way of getting on with the job were in full cry'. There was no doubt that Savige considered the 3rd Division greatly inferior to the standard of excellence achieved by the AIF he had recently left. However, he did believe in the Division's future. He confided to Hammer that the 3rd would go places once it got trained and experienced leadership. If necessary, he wrote, the men would get their experience in battle.[26]

On another occasion, and in a calmer state of mind, Savige identified what he believed had gone wrong with the 3rd Division's leadership. In a report to Headquarters First Australian Army he pointed out that the 3rd Division had provided hundreds of officers for the AIF, robbing units of their senior leaders, with the result that officers were either too young or old for their positions and were lacking in appropriate experience for service in what was now a field formation. As the best leaders found their way to the AIF, which steadily progressed, the Militia wasted away from the constant removal of its most skilled soldiers. In a sense, the Militia had fulfilled its inter-war task to serve as a nucleus for expansion. However, it did not receive the benefit of those years of preparation. Instead, it was the AIF which drew upon the Militia's experience.[27]

More specifically, Savige wrote of the problems with his officers:

- over-age and physically unfit officers filling vacancies created when more capable personnel left for the AIF;
- shortages of leaders resulting in the recalling to duty of inefficient officers;
- the expansion of the Army after mobilisation resulting in promotions that were too rapid;

- many officers lacking a realistic understanding of what was required for effective training; and
- many staff officers not being up to the task.[28]

Too many of the 3rd Division's officers, Savige believed, did not have the ability to serve in battle conditions. His solution, as he admitted to Hammer, was to 'sack right and left'.[29]

By the middle of May 1942, Savige had removed nearly 60 officers, including a brigade commander, a field regiment commander and five battalion commanders. Nor did he let up then. Officers had to perform consistently to his expectations. If they did not, he removed them and gave other men the chance to prove themselves. In January 1943, Wilton noted that Savige was visiting one of the brigades with the express purpose of straightening out some administrative problems – a euphemism for the removal of a number of officers (two lieutenant-colonels and two majors in this instance) whom Savige considered to be over-age and inefficient. After that trip, Savige noted, 'further developments later'.[30]

To expedite the 3rd Division's improvement, Savige sought to bring into the formation a leavening of experienced 2nd AIF veterans. He made his first request just three days after taking over the Division, asking for several officers from his former command, 17th Infantry Brigade, plus several more from the 6th Division's staff. His efforts were only partially successful but he persevered, and subsequent attempts brought additional AIF personnel into the 3rd Division. For example, in June 1942, the 7th Division transferred to Savige's command two majors, four lieutenants and 44 sergeants. Each battalion received four of the sergeants. In addition, Savige insisted that the sergeants were to receive consideration for commissioning at the first opportunity.[31] One of the non-commissioned officers the 3rd Division acquired from the 7th Division was Sergeant Harry Wadge, for whom the transfer was actually a return home. A former militiaman, he had joined the 3rd Division AASC in 1937. In May 1940 he had volunteered for the AIF and fought in North Africa. After rejoining the 3rd Division, Wadge again served in the formation's AASC unit, and, following Savige's instructions, received his commission while serving in New Guinea.[32]

Savige believed that hard and realistic training was required if the 3rd Division was to get ready for battle. He knew that a rigorous training program would separate the capable from the ineffective. Moreover, by driving the men hard, he would toughen them and force them to depend on each other, thereby bonding units together. To speed up the training of incoming recruits, Savige established a 'boys town', where raw soldiers received intensive training in basic skills. He also expected unit commanders to provide harder and more realistic training for their men. In a report prepared after his arrival, the 15th Brigade recorded that all its units had undergone intensive training, including exercises. In addition, the brigade held a special junior leaders' course in the evening. Each battalion also conducted special courses to improve the skills of its non-commissioned officers.[33]

To facilitate divisional level training Savige concentrated the formation at the new army camp at Bonegilla in April 1942. The 4th Infantry Brigade Group was already there, having moved to Bonegilla from Seymour in late August 1941. When Headquarters Southern Command contacted Savige about his rail transport requirements, he retorted, 'We don't want any. We'll do it on our little, flat feet'.[34] As they laboured along dusty Victorian roads under the weight of their packs and weapons, heading for a rough camp in the bush each night, the men probably did not appreciate Savige's intention.

For many young infantrymen, it was the farthest they had ever walked and some got through it only with the help of a mate who helped carry a rifle or pack. The historian of the 57th/60th Battalion described the experience as an opportunity to gain the full measure of one's comrades, something that could not be accomplished within the comfort of a camp. He continued that it locked the men together and that 'it was a great bonding process'.[35] To observers, it was clear that after each day's march the men appeared fitter, but not everyone appreciated the experience and some soldiers vented their exhaustion by directing barbed comments at their commander. One larrikin in the 37th/52nd Battalion wondered aloud whether 'our Stan got a blister on his little fat arse?' referring to Savige's use of a staff car.[36] The physical strain of the march also gave Savige a chance to assess his officers' fitness for field operations, and afterwards, he replaced several who had not met his standard.[37]

The march was not without incident. Some troops took advantage of nearby towns to slip away at night for an evening's entertainment. On one occasion, nearly an entire battalion disappeared. 30 March found the 37th Battalion bivouacked near Benalla. Although its commander, Major Charles Norton, had placed the town out of bounds, his order did not deter the men. Realising that something was amiss, Norton had the roll called at 9 p.m., only to discover that three-quarters of his command was absent without leave. He organised those remaining into arrest parties and sent them into Benalla to drive the men out of its dance halls and hotels. Those absent were fined £2. The next day, the unit resumed its march and, as they reached Bonegilla without further problems, Norton reduced the fine by half.[38]

The 3rd Division remained in Bonegilla for only a short time, during which it conducted brigade and divisional level exercises. In early May, the Allies had enjoyed their first major victory over the Japanese at the Battle of the Coral Sea, where the United States Navy turned back an enemy invasion fleet which had planned to land troops at Port Moresby to complete the conquest of New Guinea. Fearing further efforts in the region, General Sir Thomas Blamey, the Australian Army's commander, ordered the 3rd Division to Southern Queensland to protect the coastline from Brisbane to Tweed Heads from possible enemy landings. While assuming a defensive role, Savige used the time in Queensland to complete the 3rd Division's training. As Gordon L. Huggard suggested, the journey meant that 'we were on our way'. For Spielman it confirmed that the unit would eventually be going overseas.[39]

In late May 1942, trains began to leave Bonegilla for the journey north, a long, tedious trip which took most units three days. As the trains stopped at stations to feed the troops, the men could measure their progress by their meals. For example, the 57th/60th Battalion's trip began on 22 May. The route it followed was: Junee – tea, Goulburn – breakfast, Moss Vale – dinner, Gosford – tea, Taree – breakfast, Wauchope – dinner, Coff's Harbour – tea, before arriving at Casino at 2 a.m. on 25 May.[40] Signaller Robert I. Caporn remembered sitting up the whole way. On Allen G. Martin's train, the carriages were packed with men

16: Signaller Noel J. Lee working on a camp line in Queensland.
[*Courtesy* Noel J. Lee]

and equipment and the troops slept wherever there was space, including in the overhead luggage racks. Those who were drivers enjoyed more room as they travelled in their vehicles. Corporal Rayden T. Tudor spent the day in the cabin of his 3 ton truck. At night, he slept under a blanket in its cargo tray.[41]

On reaching Brisbane, Savige prepared an assessment of the 3rd Division's progress.

Men – 25 per cent have served for considerable time; 50 per cent have had three to four months training; and 25 per cent are raw recruits.
Leaders – Battalion level officers and higher are competent; young officers are enthusiastic but lacking in experience; and non-commissioned officers are fair to good
Staffs – A measurable degree of efficiency has been reached.[42]

Savige then evaluated the 3rd Division by corps. He wrote:

Artillery – Keen and hard working but practice shoots with 25 pounders are inadequate to obtain a reasonable degree of efficiency;
Engineers – Contains largest proportion of recruits in the formation. Physique is poor and training for operations inadequate.
Signals, Workshop, Maintenance – Reasonably well trained by comparison.
AASC – Reasonably good.
Medical – Good.[43]

The corps which appears to have made the most progress was the infantry. Of it he wrote, 'they are proficient in musketry, good at night work, and capable of carrying out operations in rough timbered country by day or night. They can fend for themselves'. Overall he concluded that the troops, less new recruits, were fit and hard.[44]

While perhaps his review was harsh, given the difficulties the 3rd Division had faced before his arrival, the formation had clearly made considerable progress, particularly the infantry. Savige would not let things remain as they were, however, and further intensified the training program. Most unit establishments were now approaching war levels and, as the 3rd Division deployed northward, it also moved up on the army's priority list for equipment. As a result, modern weapons and materials began to appear. For example, as the 29th Battalion prepared for the move, it received a full complement of Bren and Vickers guns, carriers and motor transport. The 57th/60th Battalion was not as fortunate but at least it arrived at Casino with fourteen Vickers machine guns, eight Bren Guns and eight Thompson sub-machine guns with which to train. Except for musketry practice, individual training was a thing of the past and unit, brigade and divisional exercises became the order of day. Savige also made physical toughening a priority, and the rough terrain of local areas such as Wild Horse Mountain featured frequently in the Division's training. Ammunition also suddenly became plentiful, allowing more live fire practice. Savige next established a jungle warfare school through which each unit progressively passed. When not on exercise, parties of men undertook operational tasks such as coast watching. For example, Lieutenant J. O. Spicer commanded a garrison of 50 men from the 37th Battalion at Burrum Heads.[45]

Savige continued to seek AIF personnel to provide the 3rd Division with a leavening of combat veterans. He also instituted an exchange program with Major-General Arthur Allen of the 7th Division, while the two formations were side by side in Queensland. Groups of 3rd Division troops did a tour of duty attached to a 7th Division unit, while a similar number of AIF personnel temporarily joined a Militia unit. The objective was to expose members of the Militia to the AIF, so that they could benefit from the veterans' more advanced skills. For example, the 3rd Division's 37th Battalion was allied with the 7th Division's 2nd/16th Battalion. Its first attachment consisted of Lieutenants Gordon Clapperton, Arthur Spruzen, John Stewart and 75 other ranks. While the exchanges lasted only a week, the battalion's historian noted that the Militia put the experience to good use in New Guinea. The scheme ended only in August 1942, when Blamey rushed the 7th Division to New Guinea to help halt the Japanese push towards Port Moresby over the Kokoda Track. After the AIF left, the 3rd Division moved further north, taking over the 7th Division's position in the Brisbane Line.[46]

A few members of the 3rd Division enjoyed a break from the rigours of training when on 7 July 1942, the Liberty Ship *Rufus King* ran hopelessly aground and split in two on South Passage Bar off Stradbroke Island near Brisbane. As the ship carried essential military supplies, the Americans decided to salvage the equipment. Douglas A. Wells was in charge of a small communications detachment provided by 3rd Division Signals. During the operation, one of his men happened to gain access to the vessel's locked bond store. For the remainder of the operation, Wells commented, the men lived happily on a supply of Canadian Club Whisky, Prince Albert Pipe tobacco and a variety of brands of cigarettes.[47]

As the crisis in New Guinea deepened, the army decided to reorganise the 3rd Division for deployment to Port Moresby. On 27 August, the Division linked the 37th and 52nd Battalions, to form the 37th/52nd Battalion, and the 58th and 59th Battalions, to form the 58th/59th Battalion. In September, Savige disbanded the 10th Infantry Brigade and redistributed its personnel. Its two battalions joined the remaining brigades. The 24th Battalion went to the 15th Infantry Brigade, while the 37th/52nd Battalion moved to the 4th Infantry Brigade. The 10th Brigade's engineers, the 10th Field Company, became the engineering element for II Australian Corps. The next change occurred on 13 October, when the 29th and 46th Battalions united to form the 29th/46th Battalion. The 3rd Division's infantry organisation now consisted of: 4th Infantry Brigade (22nd, 29th/46th and 37th/52nd Battalions) and 15th Infantry Brigade (24th, 57th/60th and 58th/59th Battalions). When the Army had re-raised the 3rd Division in 1921, there were twelve battalions on its establishment. Now there were just six.[48]

Reorganisations always have a negative effect on citizen soldiers, because the identity of their units is so closely associated with elements from outside the army. Moreover, all of the 3rd Division's units had proud lineages dating from their service with the 1st AIF on the Western Front. These were now lost to the Division and the citizen soldier tradition.

The reorganisation thus raised some concerns amongst the affected men. Lieutenant-Colonel R. P. Whalley remarked, at the 59th Battalion's disbandment review, that he regretted that the teamwork and training which had built up the unit in the last nine months was to be lost. One of his warrant officers, C. H. Howe, remembered that the event caused a degree of resentment at all levels of rank in the battalion. Jeffrey Pooler, a signaller in the 58th Battalion's headquarters company, commented that the announcement of the amalgamation came as a surprise, and no-one was happy with the news. John Satori thought that it was regrettable the army did not see fit to preserve the traditions of his 52nd Battalion.

Yet for others the change caused little trauma. Private Royce G. Ellis went on a short leave as part of the 58th Battalion and returned to discover that in his absence he had become a member of the 58th/59th Battalion. However, he did not notice any friction between the two groups and everything carried on as usual. While initially there was an adverse effect on units' morale, it was not too serious and did not last very long. After all, everyone knew that there was a job to do. More important, the battalions were now either complete or nearly so. So transient was the effect of the reorganisation that Corporal David G. Taylor wrote he was 'virtually unaware that amalgamation had taken place. We just had a different name'.[49]

The 3rd Division's restructuring in Queensland was in fact part of a broader army-wide initiative. As the danger in New Guinea and elsewhere became more acute, the government could not wait until entire divisions were ready for deployment to the operational theatre. Instead, the army sent formations forward piecemeal in response to the military situation, as units reached acceptable levels of combat effectiveness. In fact, even after entire divisions had gone overseas, their components continued to serve separately in numbers of theatres under many different commands. For example, when the Headquarters 3rd Division took control of the Wau-Salamaua area in April 1943, most of the units under Savige's command had not been a part of the formation before the war. In fact, some of the elements were not even Australian. At the same time, other parts of the pre-war 3rd Division were serving elsewhere, under other headquarters.

The need to redefine the 3rd Division was a reflection of the difficult strategic situation in which Australia found itself after Japan's entry into the conflict. There were too many threatened points, both on the mainland and in New Guinea, for the existing structure to defend, especially as the AIF had only just returned and the Militia was not yet up to strength. However, the desperate state of the Australian Army in 1942 was an outcome of pre-war neglect and government failure to improve the condition of the Militia during the two years of war which had preceded Japan's intervention. Now wartime necessity required the army to build a new 3rd Division, a fact which greatly changed the character of the formation. The result was that, when the 3rd Division entered combat for the first time in New Guinea, its new composition only distantly resembled its pre-war character.

In December 1942, drafts of men from the 3rd Division headed south for brief periods of leave. It was clear to the members of the division what this meant. They were about to move overseas. The 3rd Division was as ready as training in Australia could make it. The next challenge was combat. The men of the 3rd Division would soon get their chance to fulfil the ultimate duty of the citizen soldier – the defence of the nation in war.

Kokoda

The first element of the 3rd Division to serve in New Guinea was the 39th Battalion. Although it arrived in Port Moresby a year before the rest of the formation, and never served under Savige's command, the battalion's association with the 3rd Division was strong. In late 1941, the army realised that its defences at Port Moresby were too weak and decided to raise a Militia brigade to garrison Papua. The army designated the new brigade the 30th and allocated to it three hastily formed battalions – the 39th from Victoria, the 53rd from New South Wales and the 49th from Queensland. Recruitment for the new formation began in early October 1941.[50]

The 39th Battalion was an original battalion of the 3rd Division, and had served on the Western Front in World War I. In 1921, when the Army re-raised the Citizen Forces, the 3rd Division received the 39th Battalion. It served with the Division throughout the inter-war period, although at times being linked with other battalions. In fact, when the army decided to raise a 39th Battalion for service in Port Moresby, a unit of that designation already existed, linked with the 24th Battalion in the 3rd Division. However, rather than simply unlink the two battalions, the army chose to raise an entirely new 39th, redesignating the 24th/39th Battalion as the 24th Battalion. For a brief period, there were two 39th Battalions on the army's order of battle.

Although technically a new unit, the 39th Battalion remained true to its ancestry. Its commander, Lieutenant-Colonel H. M. Conran, actively encouraged 3rd Division members to join the unit. On 2 October he visited Seymour and interviewed 3rd Division officers, a task he also undertook with the 4th Infantry and 2nd Cavalry Divisions. On 10 October, nine officers and 523 other ranks from the 3rd Division marched into the 39th Battalion's assembly point at Caulfield Racecourse, before moving to a training camp at Darley, Victoria. One group included an entire platoon of D Company 24th/39th Battalion. The next day, a further seven officers and 400 other ranks arrived from the 4th Infantry and 2nd Cavalry Divisions. Thus the majority of leadership and members came from the 3rd Division. Even after the 39th Battalion reached New Guinea, its association with the 3rd Division continued. For example, on 23 January 1942, Headquarters 15th Brigade received instructions to provide an officer and 39 other ranks for the 39th Battalion's 3rd Reinforcement.[51]

The Japanese attack on Pearl Harbor accelerated the 39th Battalion's transfer to New Guinea. On 27 December 1941, its barely trained ranks sailed from Sydney for Port Moresby. However, instead of undergoing an intensive training program upon arrival, they spent most of their time serving as labourers at the harbour or helping to construct base facilities. For example, in January 1942, the battalion spent seventeen days unloading

MAP 5: New Guinea Area of Operations

cargo at the wharf. Other details repaired aeroplane hangars, dug trenches and built pillboxes. In fact, the only training the 39th Battalion managed during the month was a few reconnaissance patrols. Throughout March, April, May and June, the 39th Battalion's primary role continued to be that of a labour battalion. The little training the men received was limited to the level of individual weapons practice and an occasional company exercise.[52]

The 39th Battalion's role changed suddenly, however. After the enemy's defeat in the Battle of the Coral Sea, army headquarters suspected that the Japanese would try to capture Port Moresby by overland assault, via a gap in the Owen Stanley Range at Kokoda. At the beginning of July, the battalion's forward element began the trek up the infamous Kokoda Track, the area's only line of communication. The track was actually little more than a narrow footpath hacked through the dense jungle up into the towering Owen Stanley Range. The Sergeant Major for A Company, G. W. Mowat, recorded the arduous passage in his diary. With the men fresh and excited, the first stage to Uberi passed easily enough.

The next day, Mowat commented on the magnificent scenery but also noted the steep hills he climbed and the numerous rivers he crossed. On 27 July, the track had become slippery and in places Mowat had to crawl on his hands and knees over an endless number of hills. More hills followed on the 28th and, at times, he believed that he was climbing almost perpendicularly as he slithered higher only to slide down the reverse slope. The 29th was described simply as extra hard, as he scaled a mountain only to descend the other side. The 30th was another exhausting climb as he crossed the Owen Stanley Range at 7300 feet. The next two days were easy by comparison, although he had to cross many steep hills and swiftly flowing rivers. On 1 August, Mowat arrived at Deniki and took up his position.[53]

By 14 July, B Company, the 39th Battalion's lead element, had reached Kokoda and dug in. The Japanese landed at Gona on New Guinea's north coast on 21 July and advanced inland. Contact was made on 25 July. Thus, it was the Militia which began one of the Australian Army's most heroic battles. The 39th Battalion was heavily outnumbered and had no choice but to give way. But instead of racing back down the Kokoda Track, the 39th Battalion forced the Japanese to fight for every yard, and delayed their advance sufficiently for the Australian Army to rush more troops to the area. While the 39th Battalion absorbed the brunt of the assault, elements of the 21st Brigade AIF gradually arrived to help. At Isurava, the Militia dug in, intending to stand and fight until relief arrived. Joined by the 2nd/14th Battalion, the 39th Battalion continued to fight. Even the walking wounded stayed in the line, refusing to leave. During the height of the struggle, the Militia earned the respect of one commentator from the 2nd/16th Battalion, who observed, 'When I saw those poor bastards, tottering on their bleeding, swollen feet, turn round and go straight back ... I knew they were good'.[54] Finally, on 5 September, the 2nd/27th Battalion relieved the 39th Battalion. Handing over their weapons, ammunition and rations, the militiamen marched out, knowing that they had proven themselves.[55]

The 39th Battalion could now enjoy a period of well earned rest, recover its strength and absorb replacements. Meanwhile, AIF units pushed the Japanese back into their strongholds of Buna, Gona and Sananada on the coast. The enemy's resistance was fierce, and gradually Blamey had to commit more troops, including two American regiments. On 30 November, the 39th Battalion moved to Gona by plane, where it went into action again. The struggle against the enemy was so intense that, at the battle's conclusion, the 39th Battalion had been reduced to just 32 soldiers of all ranks.

The 39th then returned to Australia. However, instead of rebuilding the unit, the army broke it up, giving the men only two days' notice, and redistributed the survivors. The end came on 3 July 1943. As the rest of the 3rd Division was fighting near Wau in New Guinea, its 39th Battalion ceased to exist.

The loss of the 39th Battalion from the army's order of battle is hard to justify. At that point in the war, it was the Militia's most battle-experienced unit. It had fought with great skill and determination, and had played the leading role in saving Port Moresby during the Kokoda crisis. It had demonstrated what the Militia was capable of achieving.

Unfortunately, the army failed to recognise the value of its accomplishments and, instead of patiently rebuilding a unit which could have served as an inspiration to the force's other citizen soldiers, used most of its ranks to strengthen an AIF unit, the 2nd/2nd Battalion.[56]

The 3rd Division in New Guinea

In early 1943, the 3rd Division deployed to New Guinea. While the move confirmed that the formation was at last going into action, it also resulted in a further fracturing of the formation's organisation. Headquarters 3rd Division, with the 15th Brigade Group, went to Port Moresby, while the 4th Infantry Brigade Group, under the command of Brigadier C. R. V. Edgar, with parts of the 3rd Division's support forces, sailed instead for Milne Bay. The destination of the 10th Field Company was Buna and Morobe, on New Guinea's north coast, where the engineers provided corps level support to these growing bases. As a result of those deployments, for the rest of the war, the 4th Brigade Group and the 10th Field Company did not again serve under the 3rd Division's command.

Although the 4th Brigade was lost to the 3rd Division, it is still appropriate to outline the highlights of the rest of its wartime service. Headquarters 4th Brigade, accompanied by the 37th/52nd Battalion, arrived on 25 February 1943 at Milne Bay, where it came under the control of the 5th Division, under the command of Major-General Edward J. Milford. The rest of 4th Brigade followed over the next few weeks. The 4th Brigade's arrival allowed the army to relieve the 7th Infantry Brigade, an experienced Militia formation which had helped repel the Japanese landing the previous August. The other formation at Milne Bay was the 29th Infantry Brigade but this was also soon to leave, to join – oddly enough – the 3rd Division.

The army chose Milne Bay as the 4th Brigade's first assignment because, by that time, it was a relatively quiet area in which to introduce green troops to combat conditions. The formation's primary task was to provide base security: manning the area's static defences, patrolling the surrounding jungle to control the outlying region, and providing work parties for construction details. The last duty was the brigade's most onerous, as the Australian and American airforces were in the process of transforming Milne Bay into a major base. Still, while generally quiet, Milne Bay was not quite a backwater. Enemy stragglers remained in the area and Lawrence Armstrong, of the 29th/46th Battalion, remembers a couple of skirmishes while on patrol. Air raids were also common. The 29th/46th Battalion suffered its first combat casualties on 22 March, when bombs from Japanese planes hit its headquarters, killing Captain Norm Hirons and Lieutenant George Flynn.[57]

The passage across the Coral Sea was generally uneventful. John C. Thomas made the crossing in the *Duntroon* with the 3rd Division Signals. Harry Finn sailed on the *Taroona*, which before the war had served as a ferry between Melbourne and Tasmania. He remembers the sea being rough and the ship rolling alarmingly, so that many soldiers spent the passage seasick. The 58th/59th Battalion sailed from Brisbane on 2 March and arrived in Port Moresby on the 7th, after a brief stop in Townsville. The 3rd Division's first home in New Guinea was Shrapnel Valley, near the American bomber airfields at Port Moresby, so named because of the great quantity of shrapnel that rained down from burst anti-aircraft shells fired at Japanese planes during air raids. The 3rd Division's tenure there was brief, however. Its real destination was Wau, where it would take charge of operations on the Lae-Salamaua Front.[58]

Wau lies approximately 150 miles by air on the other side of the towering Owen Stanley Range which bisects the island. No roads pierced the imposing terrain and the only access was by air. Owing to the mountains' great height, the only place where transport planes could surmount the range was a narrow pass, known as the Gap, where the altitude was only 7000–10,000 feet. The passage was hazardous and cloud often filled the Gap, reducing visibility to zero. When this happened, nothing could reach Wau until the situation improved. Flying conditions were so difficult that it became standard practice for the airforce to first send out a reconnaissance plane to check the situation at the Gap before allowing the transports to take off from Port Moresby.

Engineers were building a jeep track over the Owen Stanley Range, but the Bulldog-Wau road did not open until late August 1943, near the campaign's end. The need to rely on air transport, and the unreliability of passage through the Gap, created tremendous limitations on the 3rd Division's logistics and helped to determine the nature of the campaign. The 3rd Division's logistical problems only eased when Savige established a sea-borne line of communications through Nassau Bay.[59]

Since April 1942, the area of operations where the 3rd Division was to take over had been under the control of an organisation the army called 'Kanga Force'. When first raised, it consisted of a local citizen soldier unit, the New Guinea Volunteer Rifles, and the AIF's 2nd/5th Independent Company. Its purpose was to observe the enemy dispositions and strength around Lae, and to undertake offensive operations at profitable opportunities. In January 1943, increased enemy activity led to Kanga Force's reinforcement by the 17th Brigade under Brigadier Murray Moten. The arrival of these AIF veterans of Greece and Syria proved timely. At the end of January, the enemy made a determined, and nearly successful, effort to seize Wau and its vital airstrip. The Japanese pushed the Australian defenders back to the airstrip's perimeter before being repulsed. The Japanese came so close to success that some reinforcements, urgently flown in from Port Moresby, went into action as soon as they de-planed. The 3rd Division's insertion into the theatre was to assure the defence of Wau, while giving the Australians the ability to take the offensive.

One of the first 3rd Division soldiers into Wau was Lieutenant Geoff Waters, the formation's Intelligence Officer. Savige sent him ahead to develop the intelligence picture and study the information Kanga Force had collected. At the beginning of April, Savige himself spent ten days in the area, examining the ground and renewing acquaintance with comrades from the Middle East. Parties of officers and other ranks from the 3rd

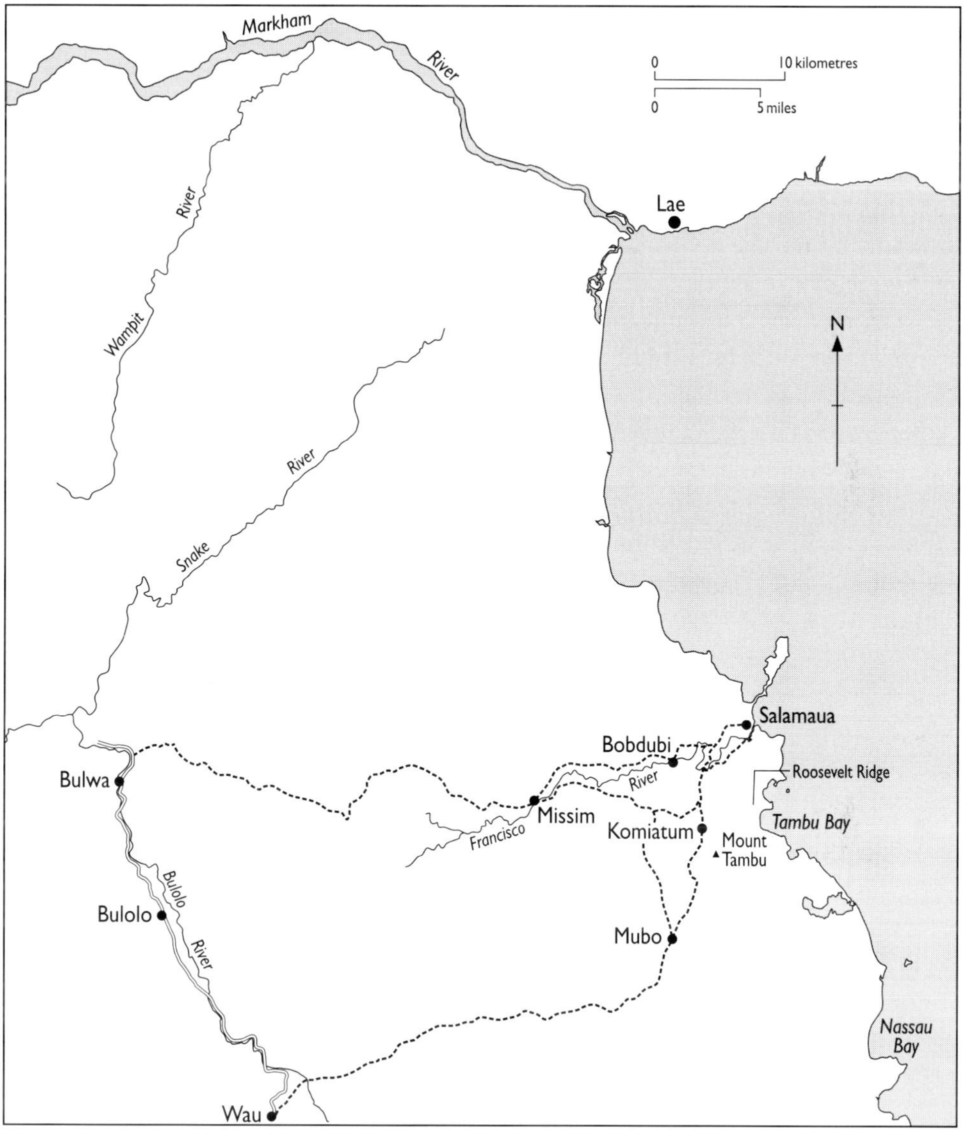

MAP 6: Lae-Salamaua Area of Operations

Division's units also rotated forward. They were attached to various Kanga Force units to gain battle experience and learn all they could about their future operational theatre. For example, a team led by Major N. M. Walker of the 24th Battalion spent two weeks with the 2nd/5th Battalion and 2nd/3rd Independent Company. The 58th/59th Battalion sent Major H. M. Heward, with a party of eight officers and five sergeants, forward on 20 March for a similar inspection.[60]

One soldier determined not to miss the action was Garry A. McGuire, a signaller with 3rd Division Signals. He had not sailed with his unit on the *Duntroon* because of emergency surgery for removal of his appendix. On recovering, he was posted to a staging camp, but, not one to wait around, he slipped away and sneaked aboard the *Duntroon*, about to make another run to New Guinea. At Port Moresby, he again stowed away, this time in a DC-3 bound for Wau. After take-off, the pilot realised that the plane was overweight but, instead of returning to base to off-load McGuire, he simply told his passengers that they had to stay in the plane's tail, to keep the nose up and gain the altitude needed to traverse the Gap. On arriving at Wau, McGuire caught a jeep to Bulolo where, saying nothing, he quietly joined the signals camp. His unexpected arrival did not go unnoticed however, and, a few days later, he was ordered to report to Headquarters 3rd Division to explain. As his adventures had brought him to the front, rather than in the other direction, he was simply told to carry on.[61]

Savige took over in Wau on 23 April 1943, at which point Kanga Force ceased to exist and its units came under the control of Headquarters 3rd Division. The majority of units under Savige's command were AIF, an unusual distinction for a Militia formation. Kanga Force's principal combat component had been Moten's 17th Brigade, with which Savige was familiar, having previously led it in the Mediterranean.

Savige's command now consisted of:

3rd Division Headquarters
15th Brigade Headquarters
24th Battalion
3rd Division Signals
3rd Division AASC.

The Kanga Force elements absorbed by the 3rd Division included:

17th Brigade Headquarters
2nd/5th, 2nd/6th and 2nd/7th Battalions
2nd/3rd, 2nd/5th and 2nd/7th Independent Companies
A Battery 2nd/1st Field Regiment
Section 2nd/1st Mountain Battery
156th Light Anti-Aircraft Battery
2nd/1st Machine Gun Battalion
Composite Carrier Company
2nd/8th, 2nd/14th and 2nd/16th Field Companies
2nd/2nd Field Ambulance
2nd/2nd Supply Company
2nd/46th Light Aid Detachment
Detachment ANGAU
Detachment Ordnance Depot
4th Army Co-operation Squadron RAAF
Detachment United States Army Air Support Control for 5th Air Force.[62]

The 15th Brigade's other battalions remained temporarily in Port Moresby, attached to the 11th Division. 58th/59th Battalion moved to Wau at the end of May. The 57th/60th Battalion did not enter the theatre until mid-June. The 15th Brigade's commander was Brigadier F. Hosking. On 3 July 1943 command of the brigade passed to Brigadier H. H. Hammer.The 2nd/5th and 2nd/7th Independent Companies were due for relief and soon left Savige's command.

The 3rd Division's area of operations stretched across a vast arc of New Guinea jungle. Wau was its administrative base and sole supply hub, at least until the occupation of Nassau Bay. Savige first established his headquarters at Bulolo. It had an airstrip suitable for light planes and was connected to Wau by jeep track. After the opening of communications via the sea, Savige relocated his headquarters to the coast at Tambu Bay. He concentrated the 17th Brigade against Mubo and the Komiatum Ridge, in the enemy's centre. The 15th Brigade, supported by the 2nd/3rd Independent Company, moved into position on the AIF's left, facing the enemy's defences along Bobdubi Ridge. The 58th/59th Battalion entered the line in late May. Savige gave the 15th Brigade's second battalion, the 24th, the tasks of defending the Wampit Valley against Japanese penetration towards Bulolo and protecting the small airstrips at Bulwa and Zenag. In addition, the 24th

17: Panoramic view of terrain around Salamaua.
[*Courtesy* AWM 055619]

Battalion was to patrol vigorously and establish bases all the way to the Markham River. The 57th/60th Battalion, the 15th Brigade's third battalion, would not participate in the assault on Salamaua but instead deployed to Tsili Tsili, a new airbase the Americans were developing on the Watut River. Its primary task was to defend the sector, and its advance elements arrived on 16 June. The 57th/60th Battalion was so distant from its parent brigade that the Headquarters 3rd Division controlled its operations directly.

During the second half of the campaign, Savige received considerable reinforcements. On the night of 29-30 June, the United States' 162nd Infantry Regiment, under the command of Colonel Archibald MacKechnie, came ashore at Nassau Bay, securing the 3rd Division's long awaited sea link. The regiment's arrival also meant that the 3rd Division now commanded American troops. Savige assigned them to the southern face of the enemy's defensive perimeter, principally along Roosevelt Ridge. On 6 August, the Australian 29th Infantry Brigade began to arrive at Tambu Bay, further augmenting Savige's strength. Brigadier R. F. Monaghan's outfit contained the 15th, 42nd and 47th Battalions. Savige inserted it between the 162nd Regiment and the 17th Brigade. Its area of operations faced Davidson Ridge and Mount Tambu. At the end of August, Headquarters New Guinea Force withdrew most of the 17th Brigade, although the 2nd/7th Battalion and 2nd/3rd Independent Company remained in the line, attached to 15th Brigade until after Salamaua's capture.[63]

At its peak strength, Headquarters 3rd Division controlled a force of three Australian infantry brigades, an American infantry regiment and various supporting arms. From one end to the other, Savige's command covered a front line that measured approximately 75 miles by air, but was actually at least twice as far on the ground. Moreover, at that time, the Lae-Salamaua Front was the Allies' most important combat zone in New Guinea, and the primary area of contest between Australian and Japanese ground forces. It was not by any means a garrison assignment or back-water posting. Instead, the Division had moved into the front line, and the location of Australia's next offensive in the South West Pacific Area. In fact, the 3rd Division had assumed the lead in Allied operations in New Guinea. This represented an unprecedented degree of responsibility for a divisional headquarters.

Lieutenant-General Iven Mackay, acting New Guinea Force commander for Lieutenant-General Edmund Herring, gave Savige his instructions on 20 April. He told Savige that he was to defend the airfields at Wau, Bulolo, Bulwa and Zenag, deny the Japanese access to the Wau-Bulolo area, and prevent the enemy from operating south of the Markham River, particularly in the Wampit and Watut River Valleys. More vaguely, he informed Savige that he was also to develop the area around Wau into an active operational zone in order to facilitate offensive operations. Lastly, Mackay warned Savige that he was not to attempt to capture Salamaua at present. On 27 May, Headquarters New Guinea Force increased the offensive aspects of the 3rd Division's orders, instructing Savige that he was to drive the enemy north of the Francisco River as soon as possible. He was also told to open a beach-head at Nassau Bay in order to establish a sea line of communications and to decrease his force's reliance on the difficult air link through the Gap.

Unknown to Savige, however, greater plans were under development at Headquarters New Guinea Force which went far beyond the 3rd Division's orders. General Douglas MacArthur, the American Commander-in-Chief of the South West Pacific Area, wanted to capture Lae as part of his larger Cartwheel operation, a series of offensives designed to seize the Huon Peninsula, as well as isolating Rabaul on New Britain. Lae was the enemy's main base in that area of New Guinea, and Salamaua, a secondary base, guarded its approaches from the south. Headquarters New Guinea Force planned to capture Lae by a combined aerial and sea assault, employing the Australian 7th and 9th Infantry Divisions. The 3rd Division's role in this operation was to be diversionary.

Amazingly, none of Savige's superiors thought to inform him of his mission's real purpose. Even by the end of May, when Savige received his revised instructions, Herring still did not bring him into the picture. In fact, Savige only found out about the impending attack on Lae just a few weeks before it took place. He did not even know that there was an additional significance to the order to open a base at Nassau Bay. A sea connection would improve the 3rd Division's communications but, more importantly, it was needed by the 9th Division as a staging point for its landing craft, as they headed for the assault on Lae's beaches. All Herring told Savige was that he was to press hard for Salamaua but not to capture the town.[64]

Unwittingly, Savige was to orchestrate a great diversion, a mission the 3rd Division fulfilled brilliantly. Through aggressive operations, bold patrols and unceasing pressure, the 3rd Division drew the enemy's attention from Lae and made it appear that Salamaua was the goal of the Australian attack. Up to the day of the landing of the 9th Division, the 3rd Division continued to attract enemy reserves to Salamaua, weakening its defences at Lae and easing the task of the AIF soldiers when they entered the battle in early September. At the same time, Savige managed to moderate the pressure his force was applying on the Japanese defenders sufficiently to avoid capturing Salamaua prematurely, which would have caused the enemy to fall back to their Lae defences, a potentially disastrous event for the 7th and 9th Divisions, which would then have faced significantly greater resistance rather than an already beaten foe.

The Environment

In the coming battle, the men of the 3rd Division faced a number of difficult obstacles which they had to overcome to achieve victory. The first, of course, was their opponent. The Japanese forces around Salamaua were experienced troops, skilled in defence, ferocious in attack and determined to prevent the Australian advance at all costs. Lae was their front line in New Guinea, and they had to hold the area if they were to recover from the setback at Buna. The incredible harshness of the New Guinea battlefield added to the Division's difficulties, greatly increasing the campaign's hardships. The men were fighting at the end of an over-extended and inadequate supply line, in an area with numerous endemic debilitating diseases, isolated by towering mountains, deep jungles and seemingly unceasing rain. What they accomplished, despite all these problems, underscores the greatness of their success.

The 3rd Division operated in an area which was amongst the most difficult theatres of war the Australian Army has ever encountered. Blamey described it as 'the foulest country' and the 3rd Division's task as 'a poisonous job'.[65] The country was extraordinarily rough. Steep-sided mountains rose into the clouds, while their bases lay in ravines through which passed swift-flowing streams. Crowning the tops of the mountains were ghostly moss forests, whose enormous trees and vines shut out the sun. All around the troops, dense vegetation and darkness limited visibility, while incessant rain ensured that clothes never dried and the ground remained sodden. John C. Thomas of 3rd Division Signals remembers the surrounding jungle as 'steamy, dank, wet, rotting and deathly quiet'.[66] Taking advantage of the terrain's natural strength, the enemy had fortified razorback ridges and mountain peaks into nearly impregnable ramparts, from which they protected the approaches to Salamaua. The terrain so dominated the campaign that even the simplest operation became an extremely difficult and exhausting exercise. The Australians soon discovered that measuring distance by miles was meaningless and instead planned their movements according to how many hours it took to traverse the intervening ground.

When the Division moved to New Guinea, Savige had believed that it had achieved a reasonable degree of training and fitness, and his soldiers were confident of their abilities. The formation had indeed trained hard, but that training assumed conditions more typical of the war's opening years. Unfortunately, New Guinea was an entirely new category of operations.

After more than 50 years, the memory of the hardships the 3rd Division endured in New Guinea remains strong in those who served there. The most consistent comments of veterans on the conditions were 'appalling', 'bloody awful', 'shocking', or that no words could describe them. William Mateer, a machine gunner in the 58th/59th Battalion described it as 'simply hell'. Spielman was no doubt correct when he wrote, 'Only a person who has served in New Guinea could understand the terrible privations we endured'.[67]

After the campaign, the veterans complained that, despite their thorough preparations, nothing they had been taught or experienced in Australia had begun to suggest the conditions they would face during the struggle for Salamaua. In 1942, no one in Australia really understood the type of training soldiers required for operations in the tropics. The formation's development in Queensland had been more a translation of the experiences gained in the Middle East to the open spaces of Australia, rather than an appreciation of the unique requirements of jungle warfare. At Caboolture, one of Jeffrey Pooler's last training experiences before embarking for Port Moresby, there was an exercise featuring his battalion's Bren gun carriers. The carrier had proved its utility in the Western Desert of Egypt but was completely unsuited for the roadless jungle of Salamaua. Fortunately, the battalion left them behind when it sailed. Upon reflection, John J. Grimes, a corporal in the 58th/59th Battalion Headquarters Company, accepted that, while his training had

18: Illustration of trail over Double Mountain.
[Illustration by Ken Budge]

toughened him up, it had not prepared him for the reality of New Guinea. While the 24th Battalion's Lieutenant Russell Mathews accepted that Savige's program made the troops physically fit and ready for combat, he also thought that the unit was inadequately prepared for jungle warfare. He believed that the Army's tropical training only became realistic with the establishment of the Jungle Training Centre at Canungra in 1943. This was too late for the 3rd Division, although the formation did go to Canungra during its refit after its return to Australia in 1944.[68]

Even the Division's specialist arms, which had achieved a particularly high degree of expertise prior to moving overseas, needed to adjust to the conditions they found in New Guinea. Spielman noted that the men of 3rd Division Signals had mastered all the intricacies of their highly technical trade, but the greatest challenge in New Guinea was making equipment, designed for the flat, arid conditions of Australia, work in an environment dominated by mountains and moisture. Their solution was to improvise new techniques for the new environment. For example, one chronic difficulty the unit's wireless operators faced was the need to keep their sets dry, a challenge in New Guinea. Eventually, moisture would win the battle, disabling the equipment until it could be thoroughly dried. The approved protocol was to take the wireless apart and allow its components to dry in the air, which was highly impractical in the New Guinea jungle. Instead, the signallers chanced upon a simpler, quicker and more effective procedure. Briefly suspending the wireless over a small petrol fire soon evaporated the moisture. The solution was highly unorthodox but, as long as the soldier was careful not to melt anything, the equipment was soon back in service.[69]

One of the functions most greatly affected by the environment was the movement of personnel and supplies through the area of operations. With few exceptions, all movement was limited to the few native tracks which wound their way through the jungle. The trails were extremely narrow, climbed up and down the mountains and were exhausting to traverse. Signaller Allen G. Martin remembers one so narrow that two people could not walk side by side, while the ground fell away nearly vertically for hundreds of feet on the downhill side. As it rained constantly, the paths were so slippery that, according to Caporn, it was 'like trying to walk on grease'. Signaller Kenneth Ballard remembers walking one track cut into a particularly steep slope and so narrow that when he rested, there was no place to sit down. All he could do was lie back against the upward slope. When the 42nd Battalion went into action for the first time, it had to scale the slopes of Mount Tambu. Bob Dillon described it as 'a stupendous climb. For the most part there seemed to be no grade at all, but just straight up and down'. He concluded that it was the hardest climb the battalion had ever experienced. Another veteran commented that the tracks were so steep that in some places the only sound you heard was your heart pumping in your chest.[70]

None of the paths were easy to negotiate, but the consensus of veterans was that the most difficult was the Missim Track. Those who experienced both it and the Kododa Track believed that the Missim was more daunting. Running from the jeep head at Power House to Missim, the trail was the main line of communication for the 15th Brigade during the battles for the Bobdubi and Komiatum Ridges. The 58th/59th Battalion advanced over the Missim Track in June, and a number of soldiers have left a record of their passage. From Power House, the first obstacle they faced was an ascent of 700 steps. Jeffrey Pooler remembers being told that this was the easiest section and, as he struggled up the steps, all he could think was 'God help us for the rest'. Setting out, Harry Finn, one of the Division's AASC soldiers supporting the 15th Brigade, looked up with awe at the almost perpendicular rise, so steep that he only made it to the top on his hands and knees. To him, it seemed to go up forever, 'until your lungs were gasping for air in the thin atmosphere [and] your legs felt like lead and your feet were hanging on the end'. John J. Grimes thought the trail was appalling. Every day, there was rain, which reduced it to a

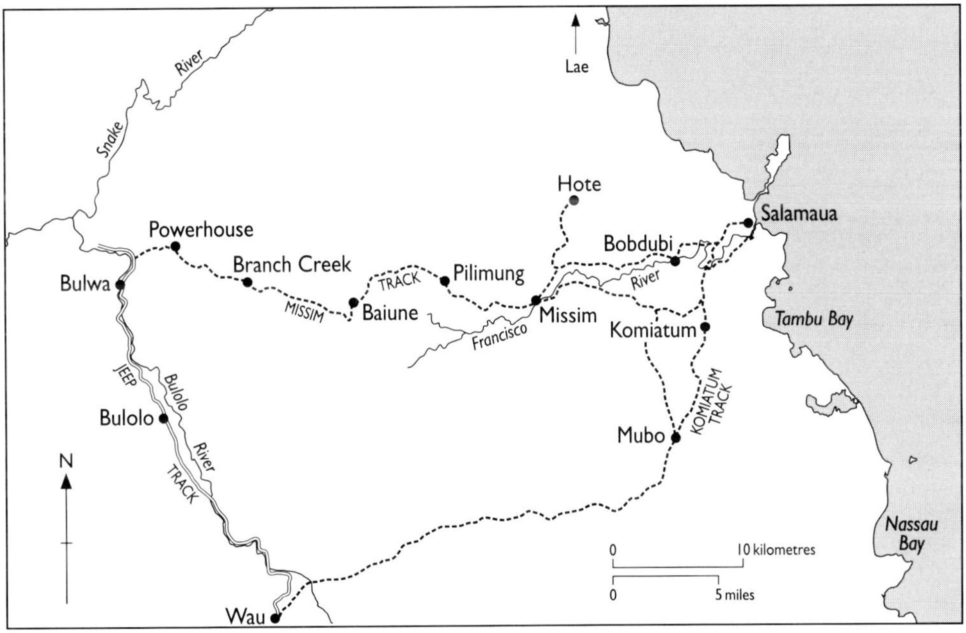

MAP 7: The Missim Track.

quagmire of mud. Some of the slopes were so steep, Grimes remembers, that the men had to get footholds in the roots of trees and help each other up.

For Corporal Colin F. Lilley, it seemed that each day was spent climbing up almost sheer ridges in mist and mud up to his thighs, only to reach a crest and slither and slide down the other side. Corporal David G. Taylor recalled not only the daily rain and mud but the cold. Although near the equator, the track crossed towering mountains, and the high altitude caused bizarre temperature contrasts. Harry Finn remembers dripping with sweat from the exertion while simultaneously shivering in the freezing air. As the temperature dropped at night, he and a mate, Bill Jackway, kept warm by huddling together under a blanket and ground sheet. Besides exhaustion, the trail presented other hazards. The mud, steep incline and undergrowth conspired to toss soldiers off the path, rewarding a mis-step with a bouncing plunge into the jungle. Finn once watched Jackway fall off the trail after catching his foot on a root. He tumbled head over heels down the precipitous slope, crashing through undergrowth until stopped by a large tree. Fortunately, he was unhurt. Each evening, the men staggered into a rest camp, ate and collapsed into sleep, only to pull themselves awake before dawn to continue the struggle over the mountains.

Without a doubt, the worst section of the Missim Trail was the part which crossed Double Mountain. Corporal Graeme T. Mills described the route as 'a nightmare' and Warrant Officer 2 C. H. Howe called it 'indescribable'. Doug W. Greenlough, a signaller with the 15th Brigade, wrote in a letter home that he would never forget the trip over Double Mountain. Pooler remembers starting off from Baiune in the early morning. The day was spent ascending to over 10,000 feet and then down for hours, only to have the trail again rise into the clouds towards the second peak. The steepness only compounded the men's labours as they pulled themselves up by tree roots or slid downwards, using their trouser seats as brakes. Pooler thought the journey would never end as he struggled up the final ridge. He did not make it to the next rest camp at Pilimung that day. Instead, collapsing exhausted, he spent the night on Double Mountain.[71]

The formation's field engineers did attempt to improve the tracks, but the 3rd Division had neither the troops nor the mechanical equipment to make more than a small dent in the job. For example, the 15th Field Company received the task of upgrading the section of the Bobdubi trail between Headquarters 15th Brigade and the Coconuts. The job required the men to cut new benches, build steps and install timber supports. The only tools available were one shovel, one machete and two axes per section. In another example, the 15th and 2nd/8th Field Companies worked together to improve the Wau-Salamaua track so that it could take jeeps. At one point, they built an 80 foot by 12 foot steel bridge with virtually no mechanical tools. As the 15th Field Company's historian observed, 'pick and shovel was the order of the day'.[72]

The region's lack of roads meant that the Division had to reexamine all its movement policies and develop new systems for the distribution of troops and, in particular, for supplies. Nowhere was it possible to operate trucks, the AASC's traditional means of distribution. Instead, for most of the campaign, all stores had to come from Port Moresby through the Gap by air, after which the lack of roads meant that everything the troops needed had to be physically carried forward from the air field. The task would have been impossible had it not been for the assistance of large numbers of indigenous people who laboured up and down the trails, bringing to the forward troops the essential supplies they needed to carry on the battle against the Japanese. For example, in the Missim area alone during the battle for Bobdubi Ridge, the 15th Brigade required more than 800 carriers to sustain the 58th/59th Battalion and 2nd/3rd Independent Company.[73]

Although the Division used porters instead of vehicles to carry supplies, the 3rd Division AASC was still responsible for organising their distribution. Units submitted supply requests, and the AASC troops organised the required stores into individual loads. These generally weighed between 30–35 pounds and might consist of a sack of canned goods, several mortar rounds, or a couple of boxes of dog biscuits, the name given to the army's hard baked biscuits which the troops received as a bread substitute. Several AASC soldiers accompanied each supply column as armed escorts. The only problems occurred when a column came under fire. Then the porters dropped their loads and disappeared into the bush.

The paucity of roads and the difficult terrain of New Guinea required the 3rd Division's AASC troops to be inventive in fulfilling their tasks. Sergeant C. R. Meilroy served in the Headquarters of the 152nd General Transport Company, a sub-unit of 3rd Division AASC. In Australia, the unit had naturally used trucks to distribute supplies to the units it supported. In New Guinea, Meilroy took to the sea and delivered supplies by small boat or barge. Lawrence Simkin helped operate a beach head, getting the barges unloaded and organising the stores into a dump. The barges always came ashore at night and the men in his unit had to work virtually in the dark. They then used carriers to move the supplies inland to units.[74]

As the tracks were so difficult, the 3rd Division relied on aerial dropping of supplies to units wherever possible. The use of forward drop zones was an important development since it effectively increased the logistic capacity of the 3rd Division's line of communications. Drop zones improved the turnaround time of porter convoys, allowing more supplies to be carried by the same number of staff. The 3rd Division AASC was responsible for operating the drop zone system. DC-3 transport planes loaded with supplies would pass through the Gap as usual but, instead of landing at Wau, would fly on to the drop zone, a small clearing carved from the jungle. These planes were nicknamed 'biscuit' or 'kai bombers'. Most of the stores fell from the plane in free fall or in bags, parachutes being rarely used.

David Tomlin was just one of many 3rd Division AASC soldiers who served at various drop zones during the campaign. His zone was near the Coconuts, in the 58th/59th Battalion's area of operations. When the Gap was open, the soldiers knew a drop would take place. The biscuit bombers arrived, flying low at an altitude of about 250 feet and travelling at approximately 250 miles per hour. The crew pushed the supplies out of the plane, as the pilot made passes over the drop zone. If the pilot was on target, Tomlin's section enjoyed a recovery rate that could reach nearly 90 per cent. However, this was exceptional, as recovery rates varied tremendously with the terrain, pilot accuracy, weather conditions and proximity to the enemy. On average, the drop zone at Ballams had a recovery rate of 40–50 per cent, and that at Skindewai 30–40 per cent.

Even if the load fell accurately on the drop zone, much of it could still be damaged by the impact with the ground. Finn, who also worked at a drop zone, noted that the square dog biscuit tins tended to end up round from bouncing on the ground. Another drop zone operator, Sergeant Allan Closter, witnessed drums of dehydrated meat and vegetables bursting as they landed. Tins of flour and sugar also tended to explode, scattering their contents, while the impact reduced loaves of bread to crumbs. Once the plane had finished its runs, the AASC and the native labourers gathered the supplies, sorted them out and stored them in their respective shelters to await distribution. Despite the losses, drop zones proved to be the most efficient means of getting supplies close to the forward units, and the Australians employed aerial dropping throughout the campaign in New Guinea as well as in Bougainville. Late in the war the army solved the impact problem by the introduction of the Storpedo, a cylindrical device with a nose cone which aborbed the force of the impact, greatly improving the delivery of fragile stores.[75]

Before the campaign began, Headquarters New Guinea Force had decided to maintain a 30-day reserve of supplies at Wau. During the campaign, the best the force's transport plane could achieve was a six-day supply, and there were times when there was only a single day's reserve at Wau. The reliance on air transport, the problem with getting planes through the Gap, and a shortage of distribution capacity in the 3rd Division's area of operations meant that the force had to set supply priorities. For obvious reasons, ammunition was the most important category, followed by medical stores. Food was further down the list, while dispatchers in Port Moresby limited mail to small letters. The men had to wait until the campaign's conclusion before they received the cakes and sweets their loved ones in Australia had so carefully baked and shipped. Soldiers also had a difficult time obtaining replacement clothing. Clarence Tonkin lost most of his spare gear during the battle for Bobdubi Ridge in June. Months passed before he received a new pair of socks and, when he went to change, those he had worn for so long disintegrated and came off in strips. Amenities were also virtually impossible to obtain. When asked about their availability, Bennett Dunstan, a Pioneer with the 58th/59th Battalion could only reply, 'You must be joking'. The Australians' only source of relief were the Salvation Army 'stalls', where the men could always count on a cup of hot coffee.[76]

One of the more frequent complaints voiced by veterans was the repetitive nature of the rations they received and the lack of fresh ingredients. The Australians never had fresh fruit, vegetables or meat except those they gathered themselves. Most troops simply had no time to supplement their food supply, and Kenneth Ogier, a signaller, was indeed fortunate to find outside his tent at Bulolo a fruitbearing paw-paw tree. Instead, men made do with food that was either dehydrated, canned, or baked so hard that it could break their teeth. Finn recalled that, whenever he opened a tin of Australian butter, it was invariably rancid, and for that reason his unit prized the New Zealand product. He also received margarine, but in the New Guinea heat it always melted to oil. The men sometimes received bread, but most of the time all they had were dog biscuits. Finn remembers that, even when soaked overnight, they were still stone hard in the morning.[77]

Field conditions required the men to accept food as it came. Once, when Finn was at Wau, the cooks served a porridge for breakfast made from biscuits and powdered milk. On the surface floated a crust of dead weevils. When the men complained to the cooks, they were told that was all there was and they could either eat it or go away. The next morning, the cooks produced the same meal and this time the men just scooped the weevils off. On the third day, the men were so hungry that they mixed the weevils in and ate the lot.[78]

Disruptions in the flow of provisions from Port Moresby exacerbated the difficulty in providing a varied diet. As a result the men had to eat the same food three times a day for extended periods. E. G. Limbrick, a Bren gunner with the 58th/59th Battalion, remembers living on dehydrated cabbage for a week. At other times, all that might be available was biscuits and bully beef.[79]

After the insertion of the 162nd Regiment, the difference between American and Australian logistic capabilities became all too apparent. The United States Army's logistic system was vast, unlike the support the Australian Army provided to its forward troops, and American troops received food in both quantity and quality of which Australian soldiers could only dream. Worse was the fact that part of the food supplies the Americans received actually came from Australian sources. The Australian government had entered into a Reciprocal Lend Lease agreement with the United States, by which Australia agreed

19: New Guinea carriers struggle down a steep slope near Salamaua with a wounded Australian soldier. [*Courtesy* AWM 015758]

to provide the American forces with all the supplies it was capable of offering within the South West Pacific Area. In return, the Australian armed forces received equipment from American factories.[80]

Some Australians were fortunate enough either to be attached to American units or to serve nearby, providing them with the opportunity to sample American fare. Douglas Wells, the signaller, worked alongside the Americans at Tsili Tsili and was astonished by the amount and variety of their rations. He remembers that the Yanks received niceties, such as ice cream and steak, which the Australians never saw in their own units.[81] On one occasion, a party of engineers from 15th Field Company volunteered to carry ammunition to an American field battery. The Yanks thanked them with an issue of coffee, chocolate and cookies. While in the American lines, the Australians observed storage sheds filled with tinned foods, clothing and equipment. After that experience, some Australians commented that they felt like transferring to the American Army. Luckily, the Americans were generous. Private T. J. Baker of the 42nd Battalion related the following story to his mother. He was with a party of Australians near an American camp. One of the Americans inquired whether they had eaten. As they had not, the Yanks shared with them their dinner of roast meat and vegetables, including fresh peas and scones. Baker wrote that, upon leaving, the Australians received 'heaps of stuff' to take with them.[82]

On other occasions, Australians acted boldly without invitation to supplement their meagre ration with American stores. When Alan Hulme was at an American camp at Nassau Bay, he saw a table piled high with blocks of chocolate, a delicacy he had not eaten since leaving Australia. The presence of an armed guard did not prevent him from acquiring a few blocks. He thought that the difference between the two armies was akin to that between a prince and a pauper. As a dispatch rider, John C. Thomas periodically had to deliver messages to American units, and remembers their generosity, enjoying their fresh meat and bread, as well as a hot shower, whenever he was at Tambu Bay. Nor was he above helping himself. Once when he was at Tambu Bay, the Japanese shelled the position. The Americans took to the safety of their fox holes and Thomas, seeing the explosions as a helpful diversion, used the opportunity to pick up some additional rations.[83]

Adding to the problems which the Australians faced was the fact that the area around Salamaua was home to a number of diseases. Malaria, dengue fever and scrub typhus were endemic, and continuous exposure to the elements increased the likelihood of outbreaks of debilitating ailments such as dysentery, tinea and tropical ulcers. Beri-beri, a vitamin B deficiency, was a problem at the beginning of the campaign. For example, a survey of the 2nd/5th Independent Company in April revealed an incidence rate of 50 per cent. Fortunately, beri-beri was relatively easy to deal with, and the addition of a small amount of Marmite to the ration quickly reduced the ailment's range. Of these diseases, malaria was the most serious drain on the 3rd Division's effectiveness. Unfortunately, as troops became carriers, they added to the infection reservoir, increasing the chance of the disease's transmission. Some soldiers fell ill more than once, and Douglas Wells noted that a trip to the hospital was inevitable. Vigorous anti-malaria measures could keep the disease under control, however. They included draining stagnant pools and the liberal spraying of Malariol. Atebrin was effective at suppressing malaria but the disease soon took hold unless the drug was taken daily. When the 3rd Division first entered the area, the malarial infection rate soared to over 4 per cent of troop strength. After a concerted effort, the formation was able to reduce this to below 1.5 per cent.[84]

The indigenous people also played an essential role in the evacuation of the seriously wounded. After treatment at an advanced dressing station, casualties who were unable to walk had to be brought by stretcher to a jeep head. This was an arduous undertaking for those carrying the wounded man, and taxing to the waning strength of the soldier. Once again, the Missim Trail over Double Mountain had the worst reputation. It took six and a half days to traverse and, on some parts, a single stretcher required eighteen carriers. It was so strenuous that even the locals could not carry for two days in a row, and only in extreme cases were the same natives who had brought supplies forward used to carry casualties back. The situation ameliorated slightly towards the end of the campaign, when the 3rd Division began operating an air shuttle service between Mubo and Wau. The Stinson 049 aircraft had a passenger capacity of one sitting and one stretcher patient and could make up to five trips a day. Its introduction reduced the four day trip from Mubo to Wau to a fifteen-minute hop.[85]

Another essential task greatly complicated by the New Guinea environment was the Division's communications. These were the responsibility of 3rd Division Signals, then under the command of Lieutenant-Colonel Lee William Fargher. The mountainous terrain and the ever-present rain made it a challenge. Perhaps the toughest job belonged to the linesmen who laid the cable for the communications network Savige required to control his widely dispersed formation. Linesmen struggled over the area's rugged terrain, clearing paths through the undergrowth while playing out wire from the heavy spools they manhandled through the jungle. Linesman Allen G. Martin remembers having first to

20: Linesmen laying line near Salamaua.
[Illustration by Noel J. Lee]

clear a path with a machete and then erect poles and cross arms to keep the copper line off the sodden ground. Whenever the enemy cut the wire, or it was broken by animals, or even friendly troops, a team of signallers had to go out to trace the break and make repairs, ever mindful of a possible ambush.[86]

If the darkness and closeness of the jungle increased the sense of isolation, at least most soldiers within the confines of a company camp or headquarters bivouac enjoyed the contact of their mates. For a few men, however, their duties denied them even this luxury, and they spent most of the campaign without any regular human contact at all. For them, the battle for Salamaua was a terribly lonely war which strained body and soul.

To keep the cable operating, 3rd Division Signals assigned individual men to patrol sections of line, a task which typically fell to the unit's dispatch riders. Normally, dispatch riders rode motorbikes, but these machines had no place around Salamaua. In New Guinea, if a message required hand delivery, a dispatch rider simply struggled over the tracks like everyone else. Fortunately, the dispatch riders had received elementary training in line maintenance before moving overseas. One such soldier was John C. Thomas. Fargher, Thomas's commanding officer, ordered him and two other dispatch riders, Gerry Britt and Colin Wallace, to maintain the signal wire on the track that ran from Power House to Missim, the vital communication link between Savige and Headquarters 15th Brigade. Thomas was responsible for the section from the Baiune Headwaters to the top of Double Mountain, without a doubt the most difficult piece of track in the theatre. Not only was the path brutal to traverse, it was also deceptive. The path up Double Mountain consisted of a series of false crests. Each time he gasped his way to the top, the view which greeted Thomas was a further ascent. Going down was just as hard, and sometimes he would fall into pools of mud which hid tree roots that delivered spine jarring jolts. Thomas commented, 'As you walked the only sounds you heard were a peculiar swishing sound made by wet trouser-legs'. He continued, 'Next was the roaring in my ears and a thumping sound which by now you knew to be your heart thumping'. During his toil there, the dense foliage blotted out the sun and sky, and weeks would pass without seeing sunshine. For two months, Thomas maintained his lonely vigil up and down Double Mountain. Despite the fear of running into the enemy, and the oppressive isolation, Thomas kept at it and made sure that the line remained in operation.[87]

On occasion the local population's assistance went far beyond portage details and included extremely hazardous operational missions. Waters, the Division's intelligence officer, recalled one instance where a villager collected critical information from the heart of the enemy's base. The native soldier had volunteered to ascertain the precise location of the enemy's 105th Division headquarters in Salamaua. He spent two nights in the town, working as a labourer unloading barges, after which he slipped away as stealthily as he had arrived. During debriefing, he picked out the two buildings making up the headquarters on a large scale photograph. A few days later, a raid by 5th Air Force bombers obliterated both structures. After the fall of Salamaua, a Japanese diary was captured by the Australians recording the destruction of the target and the causing of over 200 casualties, including 120 fatalities.[88]

The assistance the local New Guinea people provided to the Australians left a lasting impression. Even 50 years later, Grimes could still recall the care and kindness the carriers showed to his wounded mates. Robert I. Caporn, a wireless operator, owed his life to one local lad, who dived into the Yellow River and dragged him out before he was swept downstream. In Caporn's opinion the natives could not be praised highly enough. Private E. G. Limbrick, a Bren gunner, described the New Guinea people as wonderful and thought victory would have been impossible to achieve without their help. Douglas Wells, with 3rd Division Signals, also thought the locals were wonderful people and did not know how his unit would have managed without them. After the campaign ended, a report prepared by 3rd Division Signals summed up the contribution of the native carriers as magnificent.[89]

The Battle for Salamaua

When Savige took over from Moten in Wau, the 17th Brigade had already planned an attack by the 2nd/7th Battalion against The Pimple and Green Hill, two features guarding the eastern flank of the key Japanese position at Mubo. Despite air and mortar support, the AIF was unable to overcome the enemy's strong defences and, by the evening of 25 April, it was apparent that the attack had failed, although the battalion made further efforts into early May. In the battle's aftermath, Savige decided that the 3rd Division would no longer make frontal assaults against the enemy's prepared positions around Mubo. Instead, it would pursue a policy of aggressive patrolling and minor limited offensives. The Australians would flank and penetrate the enemy's defensive perimeter in order to neutralise their effectiveness by cutting their foe off from supplies, communications and reinforcements. Savige, therefore, had set the offensive pattern for the rest of the campaign. The Australians would henceforth rely on extensive patrolling to isolate and neutralise their opponent's forward positions and, when the time appeared ripe, launch short sharp attacks to dislodge the weakened Japanese from their defences.[90]

Patrolling became the 3rd Division's dominant operational activity at the tactical level, and it was one at which the Australians excelled to such an extent that they slowly took control of the jungle and denied it to the enemy. Savige ordered that 'the enemy should be constantly harassed by raids, ambushes and fighting patrols'.[91] As well as being an effective way to weaken the enemy, patrolling also made strategic sense in the terrain and environment of New Guinea. In the 3rd Division's vast area of operations, small unit actions magnified the formation's strength and allowed it to control a much larger territory than would otherwise have been possible. In contrast to World War I, the situation in front of Salamaua did not allow the 3rd Division to maintain a continuous front line. Instead, Savige dispersed his units, or even sub-units, at strategic points or along lines of operation. The most extreme deployment was perhaps the 24th Battalion, which, establishing a series of posts, thinly held a front 30 miles wide by ten miles deep. One of its platoons, the 15th, was responsible for a fifteen-mile stretch of the Markham River. Unit positions resembled isolated islands, separated from one another by deep jungle and rugged mountains. Although posts were physically separated, patrols bound the 3rd Division's disparate elements together, maintaining contact between unit positions and controlling the intervening ground. The art of patrolling was of such importance that Savige was to highlight it in a training manual on jungle warfare that he wrote in preparation for the Bougainville campaign.[92]

Every morning, and at irregular times throughout the day, each platoon sent out a contact patrol to its company headquarters, and to the adjacent position. Each company also put a standing patrol in front of its position, from dawn to dusk, to guard against a surprise attack. Contact patrols consisted generally of only two or three men, typically drawn from the unit's support arms, in order to rest the infantry for combat. Their primary mission was to control the intervening jungle and to discover any evidence of enemy presence.

To take the war to the enemy, units also frequently sent out ambush or fighting patrols. The objective of these patrols was to inflict casualties on the Japanese, interdict their lines of communication and prevent them from bringing forward supplies to their forward units. The long-term objective of these tactics was to sap the enemy's strength and morale, and convince them of the Australians' superiority. Private Royce G. Ellis, of the 58th/59th Battalion, was part of a text-book ambush. Travelling light, the members of his patrol infiltrated the enemy's perimeter and took up position on a Japanese supply trail. Shortly after their arrival they heard an enemy party approaching. When it entered the killing zone, the Australians opened fire and a number of enemy soldiers fell, wounded or dead. Ellis' party then disengaged and withdrew without loss. Signaller Jeffrey Pooler from the same battalion's headquarters company went out on a similarly successful mission. A party of seven Japanese appeared and the Australians poured fire into their ranks, cutting down three while the rest fled. Perhaps unwisely, Pooler's party then pursued the retreating enemy until it came under fire from another enemy group. Fortunately, they were able to withdraw without loss.[93] Sometimes, instead of setting an ambush, a patrol stalked its opponent. On one patrol, Corporal J. B. Leslie, of the 24th Battalion, found an enemy signal wire. His party followed the line to an enemy camp where they found eight Japanese soldiers bathing. All were killed.[94]

The 3rd Division also made extensive use of reconnaissance patrols to gather intelligence. Lieutenant E. R. Robinson took a party down the Watut River to its confluence with the Markham. Savige had briefed Robinson personally, instructing him to determine whether the Japanese had infiltrated or were patrolling the Watut Valley, and to make a detailed reconnaissance of the area to locate positions which the Australians could occupy if the enemy threatened the area. Savige was concerned that the Japanese might use the Watut Valley to strike into the 3rd Division's rear. Robinson was out for nineteen days. He examined native bridges, located fords and identified possible defensive positions and air dropping sites. He concluded that the enemy was not exploring the Watut and, moreover, reported that its garrisons along the Markham River were overly confident and careless. Savige was interested in Robinson's findings and soon acted, sending him back to Port Moresby to brief the Headquarters 5th Air Force on the area's ability to support an air base. Robinson's patrol thus provided the information which led to the development of Tsili Tsili as an American air base, which became the area of operations for the 57th/60th Battalion.[95]

These patrols were extremely risky. The Australians were deliberately putting small, lightly armed parties of men into territory either uncontrolled or in the hands of the Japanese. Occasionally, an ill-fated patrol ran into a superior force, and then the Australians were essentially on their own, inside enemy lines, until they could reach the safety of their

21: Troops of the 2nd/5th Battalion on Mount Tambu Track during the Salamaua campaign. [*Courtesy* AWM 056762]

own positions. This is precisely what happened to Private J. T. Quirk of the 58th/59th Battalion. He was part of a patrol which had established itself in an ambush position astride the Komiatum Track, the enemy's main communication line. When the enemy entered the killing zone, the Australians opened fire and a number of Japanese fell. However, the Japanese force was large and well armed, and increasing fire soon forced Quirk's party to retire.[96]

On a different patrol, the first William Mateer knew of the enemy's presence was when a burst of machine gun fire ripped into Corporal S. P. Fidler. Private A. C. White crawled forward and attempted to pull Fidler to safety, but he too was hit. Mateer and another soldier succeeded in dragging Fidler and White back, but both soon died. Mateer took charge of the patrol, fell back a short distance and organised an all-round defence. Fortunately, the Japanese failed to press their advantage and when the Australian party again advanced, the enemy had disappeared.[97]

Private Harold Hibbett was part of a patrol which ran into trouble. Drawn from A Company, 58th/59th Battalion, its orders were to interdict enemy movements along a trail near Orodubi. Hibbett remembers occupying an ambush position overlooking the junction of two tracks. In the afternoon, a strong force of Japanese appeared. Armed only with rifles, two Owens and a Bren gun, the Australians opened fire, but the enemy replied with mortar and machine gun fire. The fusillade forced the Australians from their position. Hibbert and two other men withdrew down a small track only to run into an enemy flanking party. They bolted into the jungle and, after two days, reported back to their battalion. On another occasion, a combined ten-man patrol drawn from the 24th and 58th/59th Battalions under the leadership of Corporal J. B. Leslie had manned an observation point behind the enemy's line for several days before the Japanese discovered its presence. Disaster struck when the Japanese ambushed the patrol. The seven survivors, including a stretcher case, had to scrub-bash around Japanese positions but, after three days hacking through the jungle, they regained their lines.[98]

While patrolling set the tone of the campaign, the infantry still had to force the enemy from its positions. However, Savige was determined to avoid costly frontal attacks. Instead,

22: Men of D Company, 2nd/5th Battalion cover a track within 50 yards of Japanese lines. Left to right are Corporal R. J. Jones, Corporal E. Thompson and Private C. E. Wilson. [*Courtesy* AWM 056767]

as reinforcements arrived, he extended the perimeter, stretching the enemy's defences and increasing the strain on his opponent. In addition, Savige applied pressure to the Komiatum Track, the enemy's main line of communication. If the Australians cut it the enemy's defence of Mubo, Mount Tambu and other strong points would be untenable.

After a brief pause, following the 2nd/7th Battalion's failed attack on The Pinnacle at the end of April, the 3rd Division returned to the offensive on 3 May. Patrols from the 2nd/3rd Independent Company had determined that the enemy held the Bobdubi Ridge only lightly. An opportunistic attack by these jungle veterans pushed the enemy off the ridge and they occupied the key positions at Old Vickers and the Coconuts. The seizure of these positions put Australian forces within striking distance of the Komiatum Track. Aware of the threat to their line of communications, the Japanese launched a well supported counter-attack on 14 May, pushing the Australians out of Old Vickers. The Australians tried to regain the position but were unable to dislodge the enemy. As a result of the loss of Old Vickers, the 2nd/3rd Independent Company had to abandon most of its recent gains. The Japanese now began to fortify the ridge in order to safeguard their communication link to Mubo. Bobdubi Ridge became one of the campaign's most contested positions, as the Australians strove to breach its ramparts to interdict the enemy's critical supply lines.[99]

In late May, Savige began to act on revised instructions from Headquarters New Guinea Force, to increase the tempo of the Australian operations, push the Japanese north of the Francisco River, and open a beach head at Nassau Bay. This resulted in Operation Doublet which commenced on 30 June. Savige planned a three-stage attack. First, the 162nd United States Infantry Regiment would land at Nassau Bay, supported on the landward side by elements of the 2nd/5th Battalion. On the same day, the newly arrived 58th/59th Battalion would seize Bobdubi Ridge, and then thrust southwards to secure the Komiatum area, isolating the enemy's forces around Mubo. The 24th Battalion was to carry out a diversionary manoeuvre towards the Markham Valley, and also establish an ambush position at the mouth of the Buang River to interdict the movement of enemy reinforcements. Once the Americans were in position, they would assist the 17th Brigade to clear the Japanese from their defences around Mubo.

The Americans landed just after midnight on 30 June. In the high surf, most of their boats sank, but while they lost nearly all of their stores, all the troops got ashore safely. While further waves of Americans came ashore, the covering force provided by the 2nd/5th Battalion easily repulsed a small enemy attack. By the end of the day, Savige had a secure beach head. One of the 162nd Regiment's battalions moved inland to assist the 17th Brigade, while the rest of the American force moved along the coast towards Tambu Bay.

Unfortunately, the 58th/59th Battalion's introduction to combat was not as auspicious. The Japanese largely repulsed their attack, ruining the co-ordinated aims of the offensive. However, it is doubtful that even the most experienced troops could have done better, given the circumstances of the battalion's movement into its assault position. The forward elements of the battalion moved out from Bulwa on 19 June for the approach march over the infamous Missim Track. Additional sections followed on subsequent days. Each man had to carry his own gear plus a share of the battalion's equipment and reserve ammunition. The weight made the normally strenuous march over Double Mountain a nightmare.

The final advance to the unit's start points was made during the night before the assault. The exhausted men reached their assembly areas in the early hours of 30 June, only a short while before the commencement of the attack. They had little chance to recover from their march or acquaint themselves with the terrain of the coming battle. Compounding the problems the battalion faced was that bad weather prevented most of the support bombers from arriving. The single plane that did make it through the Gap dropped its load on the wrong target. At 9 a.m. the Australians crossed their starting line. Initially, they took the Japanese by surprise, but the enemy soon recovered and poured machine gun fire into the Australians from fortified and camouflaged positions. Doggedly, the Australians pressed forward and, over the next two days, managed to secure some of their objectives. However, Old Vickers, the critical objective, remained in enemy hands and the battalion was unable to turn the enemy's flank towards Komiatum. While it had seized some terrain, the battalion had failed to realise the larger aspects of the operation. The 58th/59th Battalion had secured a foothold on the Bobdubi Ridge however and, over the next two months, its men pushed the enemy from their remaining positions, opening the way to Salamaua.[100]

On 7 July, the 17th Brigade and the 162nd Regiment began their attack on the enemy's strongholds around Mubo. Over the next seven days, they pushed the Japanese from The Pimple, Green Hill and Observation Hill. The capture of Mubo soon followed. The Japanese now fell back and re-established themselves along a line running from Roosevelt Ridge along the coast, over Tambu Mountain and then along Komiatum and Bobdubi ridges, to the banks of the Francisco River. While the 3rd Division had pushed the enemy back, the Japanese had retreated into a second formidable defensive line which blocked the way to Salamaua.

On 28 July, the 58th/59th Battalion captured Old Vickers on Bobdubi Ridge. The Japanese tried to regain the position but were beaten back after a series of attacks in which they lost at least 79 killed.

At this time, the 17th Brigade and the 162nd Regiment also began a series of operations to isolate the Japanese on Mount Tambu. After its arrival, the 29th Brigade took over from the 17th Brigade. On 11 August, the 42nd Battalion occupied Davidson Ridge on Mount Tambu's right while, on 16 August, the 2nd/6th Battalion pushed on to Komiatum Ridge. With the capture of these two positions, the Australians had virtually surrounded the enemy on Mount Tambu, separated only by a narrow ravine. Despite the fact that they were on different ridges, the 42nd and 2nd/6th Battalions were so close that they could support each other with defensive fire. When the Japanese launched a violent counter-attack against Komiatum Ridge on 15 August, it was the 42nd Battalion's fire from Davidson Ridge that annihilated the enemy. For several days, the Japanese made desperate attacks but the Australians repulsed each, with heavy losses. On 19 August, 1st Battalion, 162nd Regiment moved on to Mount Tambu without opposition. Over the next several days, the

Allies mopped up small enemy parties and stragglers. Meanwhile, over a period of two weeks the rest of the 162nd Regiment had cleared the enemy from Roosevelt Ridge.[101]

Savige was to see neither the capture of Salamaua nor the AIF's landing at Lae. As the campaign's climax drew near, Headquarters New Guinea Force rotated Headquarters 3rd Division out of the line and replaced it with Headquarters 5th Division. Headquarters 5th Division had been serving at Milne Bay, largely in a garrison capacity. The change in command took place on 23 August, when Savige handed over to Milford. The switch in divisional headquarters was part of an army policy to give all its formations experience in combat operations. However, there was another factor in Savige's replacement. Blamey believed that Savige had pressed the 3rd Division's attack in front of Salamaua too hard and had risked dislocating the Allied strategy for the seizure of Lae. In mid-August, Blamey was in Port Moresby, meeting Herring at Headquarters New Guinea Force. Blamey sent his deputy, Major-General Frank Berryman, to see Savige and report on the situation. While with Savige, Berryman informed him of his imminent removal. Savige returned to Port Moresby where he successfully exonerated himself by explaining that he had operated properly, based upon the orders he had received and the operational plans to which he had been made privy. In fact, he pointed out that he had only recently been informed of the Lae operation's existence and that the true nature of the 3rd Division's mission was to create a diversion. The defence saved Savige's military career and he returned to Australia with Headquarters 3rd Division.[102]

After enjoying a period of leave in Melbourne, and a celebratory parade through the city's streets at the head of the 17th Brigade, Savige and Headquarters 3rd Division moved to the Atherton Tablelands in Queensland in December 1943. Temporarily, the 3rd Division now consisted of the 7th, 8th and 23rd Infantry Brigades. Of those, only the 7th was to serve with the 3rd Division in its next campaign in Bougainville. In effect, the 3rd Division's role was now that of a training headquarters, and Savige's task was to disseminate the lessons learned in New Guinea to the units under his command.

After more than two years, Savige's command of the 3rd Division was soon to come to an end. In early 1942 it had been a green Militia outfit incapable of field operations against a first class opponent. As the Division's components rested after their New Guinea assignments, the change was obvious. The 3rd Division was a veteran, combat-hardened formation which had met and bested the enemy. In April 1944, promoted to Lieutenant-General, Savige left the 3rd Division and returned to New Guinea to take up a corps command. His successor was Major-General William Bridgeford. Savige however, was to renew his association with the 3rd Division in Bougainville, where he would oversee the conduct of the campaign as commander of II Australian Corps.[103]

The Campaign Continues

While the Headquarters of the 3rd Division languished in Australia, its traditional components remained in New Guinea, under the command of other headquarters. Although the command structure had changed, 3rd Division elements continued to be identified by their existing names. Thus, a confusing situation existed in which 3rd Division Signals served as the signal unit for the 5th Division. The constantly changing associations of 3rd Division Signals illustrated the shifting allegiances under which units of the 3rd Division served. After the conquest of Salamaua, 3rd Division Signals moved to Lae, where it provided communications for the Headquarters 5th Division but also became a de facto corps signal unit for the entire base area. The signallers remained in Lae until February 1944, when they embarked for Australia. Upon completion of leave, the unit reassembled at the Atherton Tablelands, where it again joined the 3rd Division. However, after retraining, its members again parted, and the signallers returned to Lae. When Australia took over responsibility for Bougainville from the Americans at the end of 1944, the 3rd Division was one of the formations sent to wage the campaign. In early October, the signallers embarked for Bougainville and rejoined the 3rd Division.[104]

The long awaited offensive against Lae began on 4 September. The 9th Division landed on beaches to the west of the town, while, on 5 September, the 7th Division began to fly into the Nadzab airdrome to the east, seized by paratroopers the previous day. To support the attack, and to confuse the enemy as to its main objective, the 24th Battalion attacked from the Wampit Valley towards the Markham River. While initially offering determined

23: Mortar crew from 57th/60th Battalion make their way down a road near Yaula, New Guinea. [*Courtesy* AWM 072263]

resistance, particularly against the 9th Division's crossing of the Busu River, the remaining Japanese defenders fled what had become an untenable position. The Australians occupied Lae on 16 September. Meanwhile, Salamaua fell to the 5th Division on 11 September. The 3rd Division had done its job. The Japanese had gradually committed their reserves to the defence of Salamaua and, when the attack on Lae began, could no longer repel the Australian assault.

After the enemy's collapse, the 15th Brigade enjoyed a well earned rest. For most of its ranks, it was the first time they were out of the dank, jungle-clad mountains surrounding Salamaua since their arrival in New Guinea five months earlier. The 58th/59th Battalion in particular had earned a respite, having been in continuous daily contact with the enemy for 77 days. Seventy-eight members of the battalion had died and a further 140 had been wounded. The pause in Salamaua at last allowed the brigade's postal staff to distribute the men's parcel mail which during the battle, had gathered at the rear. When Sergeant Les Jones, of the 15th Brigade's postal section, arrived at Tambu Bay he found a small mountain of items awaiting collection. He returned to Salamaua in a heavily laden barge, and the men eagerly consumed the cakes and other luxuries despatched to them months ago. Although now quite stale, the treats were a vast improvement over the bland, repetitive diet they had endured for so long.[105]

Soon afterwards, the 15th Brigade moved back to Port Moresby and then to the Donadabu Highlands, behind the city. The men had hoped that their relief would include a trip back to Australia, but they were to be disappointed. Instead of further leave, in early January 1944 the 15th Brigade returned to the fighting line, to Dumpu in the Ramu Valley, under the command of the 7th Division. The brigade then supported the clearing of the Ramu Valley and the advance on Madang. Lieutenant G. H. Atkinson, leading a patrol of the 57th/60th Battalion, captured Madang on 24 April 1944.[106]

As the Australians advanced on Lae, the time for the 4th Brigade's blooding had arrived. Edgar's brigade spent nearly all of the Lae-Salamaua Campaign in Milne Bay, largely in a garrison role. On 10 September it came ashore at the 9th Division's beachhead and rejoined the 5th Division. Initially, the brigade's task was to help provide security around Lae, including sending patrols into the surrounding hills to track down enemy parties. However, Headquarters New Guinea Force soon detached it from the 5th Division and placed it under the command of the 9th Infantry Division. One of its battalions, the 22nd, had been attached to the AIF formation since shortly after its arrival at Lae. The 4th Brigade was to assist the 9th Division in its next offensive, the clearing of the coast of the Huon Peninsula as far as Sio. The attack began on 5 December with the 4th Brigade in the lead. A fortnight later, having reached Fortification Point, the 4th Brigade moved into support and the 20th Brigade took over the attack.

In the New Year the 4th Brigade reverted to the 5th Division's control and participated in the pursuit of the Japanese along the coast to Madang, after the victory of the 9th Division at Sattelberg. The 4th Brigade's last operation in the New Guinea campaign was the capture of Karkar Island in June. The landing was anti-climactic, as the island proved to be unoccupied.[107]

The occupation of Madang brought to an end the 4th and 15th Brigades' operations in New Guinea. In August 1944, after a year and a half's absence from Australia, the brigades rotated home. Their relief followed a familiar pattern. The men were reintroduced to civilisation on the voyage across the Coral Sea. Alan A. Hulme, from the 3rd Division Signals Headquarters Company, made the passage on *Katoomba*. He thought the crew was playing a practical joke on him when, during his first meal on board in the non-commissioned officers' mess, the stewards presented him with a menu card. After the deprivations of New Guinea, many men were no longer accustomed to eating fresh food and soon came down with gastric disorders. One soldier was so determined not to miss the voyage home that he neglected to report to the medical officer at Port Moresby when he came down with dengue fever. He took a few aspirin, and stumbled on to the ship, while his mates carried his gear on board for him. He spent the voyage recovering, but he got back to Australia.[108]

Upon reaching Australia, many men neglected to continue taking their Atebrin, with the result that they spent part of their leave in hospital recovering from attacks of malaria. Hulme, for example, was hospitalised at Heidelberg, while Thomas, the dispatch rider who had maintained the signal line over Double Mountain, spent part of his leave in a Brisbane hospital. At one point, the 42nd Battalion had more men in hospital with malaria than in its Brisbane camp. During August, the peak month, the number of malarial evacuations by the battalion reached an astonishing 269.

At some point during their time in Australia, units generally participated in a parade. The 42nd Battalion joined the rest of the Queensland-based 29th Brigade on 26 September for its march through central Brisbane.[109] On 13 October, the 4th and 15th Brigades, temporarily together again in Melbourne, participated in a March of Honour through the streets of the 3rd Division's home base. The Minister for the Army, Francis Forde, was in attendance. Reviewing the two brigades, he appropriately commented, 'These men have done all that has been asked of them. They have been asked to do more than we could rightly expect'.[110]

The men enjoyed their well earned leave before reassembling for retraining and refitting in Queensland. The 4th Brigade rejoined the 5th Division on the outskirts of Brisbane, while the 15th moved to the camps on the Atherton Tableland, where it again came under the command of the Headquarters 3rd Division.

Bougainville

For most of the 3rd Division, the sojourn in Australia was all too brief. However, the purpose of the 3rd Division's next deployment was to be different from the essential nature of their fighting in New Guinea. If, during 1943, the 3rd Division had fought at the war's front line, its final campaign was to take place at the rear, far removed from the

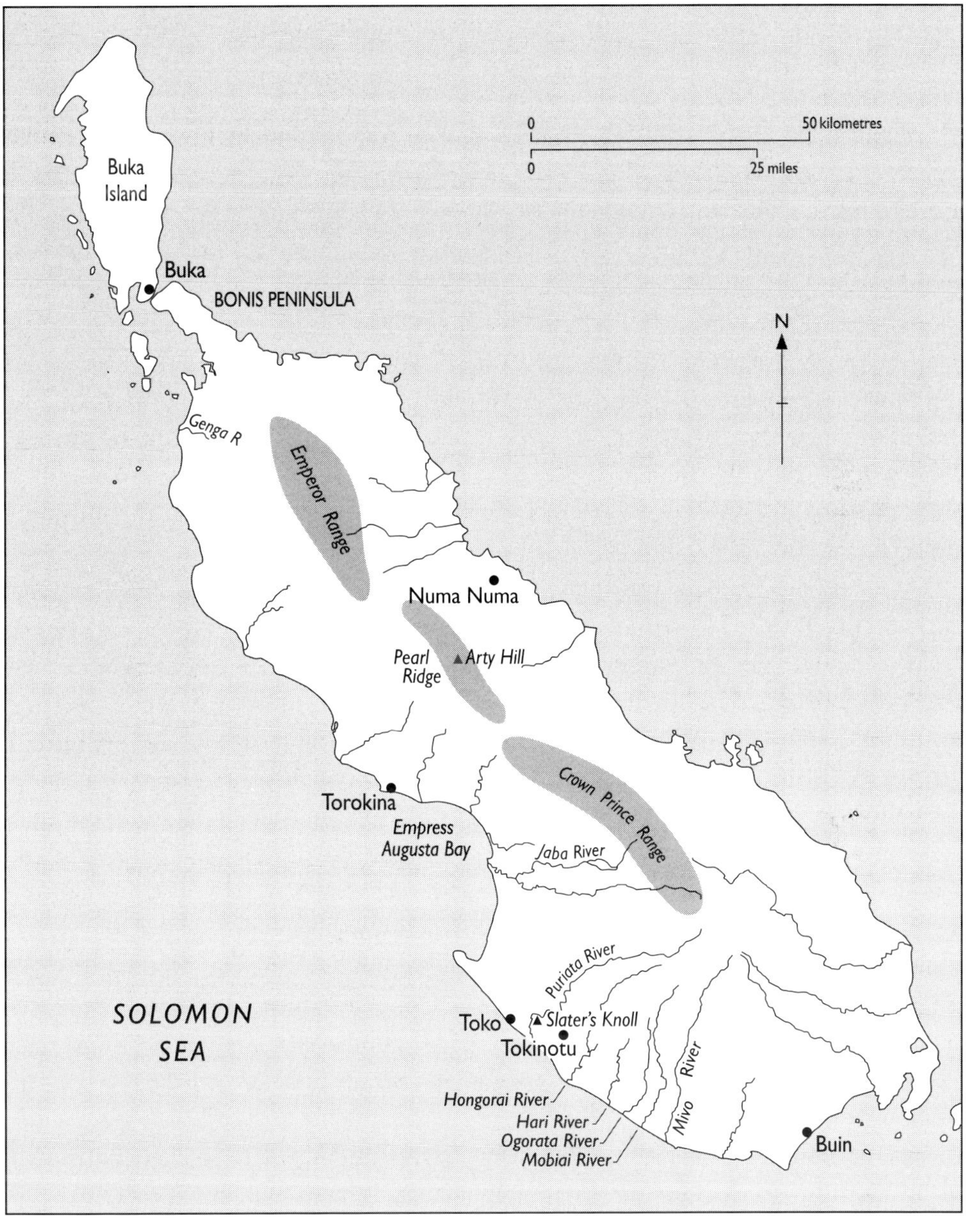

MAP 8: Bougainville Area of Operations

United States' march towards Japan and without any influence on the conflict's outcome. By now the war's balance had shifted decisively in the Allies' favour. The fact that United States naval forces roamed freely across the Pacific confirmed that there was little doubt that Japan was losing the war. Although, in the South West Pacific Area, considerable numbers of enemy troops remained on New Guinea, and on the other major islands, New Britain and Bougainville, MacArthur's focus was solely on the Philippines. His enormous ego prevented him from sharing that honour with his junior ally, and instead, he would marginalise the role of the Australian troops under his command for the remainder of the war.

In mid-1944, MacArthur had under his control eighteen American and six Australian infantry divisions. At that time, six of his American divisions served in what were largely garrison roles, where they contained by-passed enemy troops on New Guinea, New Britain and Bougainville. MacArthur wanted these divisions relieved by Australians so that he could use the Americans in the forthcoming Philippines campaign he planned to commence in October. As the Australian Prime Minister, John Curtin, did not oppose his request, Blamey had no choice but to implement MacArthur's wishes.[111]

Blamey assigned the 3rd Division to the II Australian corps, together with the 11th and 23rd Brigades, and ordered Savige, the corps' commander, to Bougainville where he took over from the United States Army's XIV Corps. Blamey's other divisions also spent the remainder of the war fighting in areas of little strategic significance. The Australian 5th Division was to go to New Britain, while the 6th Division and the 8th Infantry Brigade continued the reconquest of New Guinea. The 7th and 9th Divisions, Australia's remaining combat-capable formations, invaded Borneo. With Curtin's assent, MacArthur had assured that Australian soldiers continued to fight and die until Japan's surrender, but that their sacrifice was without meaning.

Although MacArthur determined the Australian Army's location and strategic role, it was up to Blamey to decide the operational objectives and intensity of combat that his force directed against the Japanese. Since their arrival on Bougainville, the Americans had assumed a defensive attitude around their base at Torokina, and had left nearly all of the island to the enemy. The Japanese adopted the same policy after American firepower bloodily repulsed their attempt to push the Americans into the sea in March 1944. After that effort, both sides settled into a comfortable live and let live system of mutual accommodation, where they defended their own lines but did not undertake any significant offensive action against their opponent.

Blamey decided that this was unacceptable. Instead of allowing the Japanese free reign over nearly all of Bougainville, he wanted the Australians to take the offensive and contain the enemy in two small enclaves at the island's extremes. He ordered Savige to strike out from the encampment at Torokina, on Empress Augusta Bay, and push the enemy towards Buin in the south, across the central range in the centre, and the Bonis Peninsula in the north. Torokina was ideally suited to support these drives, as the Americans had built up a well developed base infrastructure, and it lay on the island's west coast, roughly at the island's mid-point. The Japanese naturally resisted these efforts and undertook offensive action of their own. The 3rd Division, therefore, once again found itself fighting against a skilled and determined opponent who resisted to the bitter end.

The need for the Bougainville campaign and Australia's other operations in the war's final year have engendered considerable controversy by both contemporaries and later scholarly observers.[112] Even in 1945, commentators critised Blamey's actions, despite his justification on the basis of military necessity.[113] More recently, Peter Charlton described these offensives as attacks upon 'self-supporting prisoners of war'[114] forced by the Australians to fight, and John Hetherington has questioned whether these battles 'served any war-winning purpose whatever'.[115]

Of the protagonists, MacArthur's motives are the easiest to understand. Australian troops were on Bougainville and New Britain at his insistence, as he refused to share the triumph of his return to the Philippines with the troops of another nation. Moreover, he had determined the allocation of troops and overruled Blamey's suggestion that all these operations required were fairly modest garrisons. MacArthur took the unusual step of insisting that Blamey send an entire corps to Torokina. This forced Australia to commit a large part of its army to a backwater operation, and gave MacArthur a convenient excuse for not including his ally in his future plans.[116]

However, while scholars may question MacArthur's motives, it was Blamey who set the campaign's operational policy and decided to end the 'live and let live' policy the Americans and the Japanese had accepted. Blamey realised that Australia did not have the ability to mount major, American-style offensives, especially as MacArthur took most of the South West Pacific Area's shipping and support elements with him to the Philippines. He was still determined that his soldiers would be aggressive however. In particular, he was determined that Australian ground forces, after their many achievements earlier in the war, would not become 'a second-line army reserved for second-line tasks'.[117] He decided that the Australians would follow a policy of intensive patrolling and limited attacks, similar to those they had used to such good effect in New Guinea. Besides causing casualties and controlling the jungle, the Australians were to seek out and destroy the enemy's jungle gardens, from which they derived their food supplies. Instead of dictating a passive occupation, Blamey's policy would provoke the Japanese into a spirited defence.

Ultimately however, it is Curtin who is the most culpable. As the political leader of an allied force, he had the right and duty to object to MacArthur's plans, and to demand a more significant role for his nation. However, as he had so often in the past, the Prime Minister acquiesced and allowed MacArthur to dictate Australian military policy.

Moreover, for Curtin, and his successor as Prime Minister, (Joseph) Ben Chifley, there was political benefit in having Australian troops engaged on active operations at the war's end. The fact that Australia had fought until victory would strengthen its claim to a role in any settlement and would possibly influence its allies in the formulation of the post-war strategic environment. The Minister for the Army, Francis Forde, observed that it was of great importance for Australia to be a part of the drive to defeat Japan, and in particular to maintain the gratitude of the United States, to foster American-Australian relations in the post-war period.

A further justification was that, technically, the soldiers fighting these campaigns were liberating Australian territory that the nation had acquired under a League of Nations mandate, as part of its spoils after World War I. More important, however, Curtin was aware of the army's plans for the Bougainville campaign and he could have demanded that Blamey change the operation's style. Instead, he chose to allow the battle to proceed, resulting in the sacrifice to the nation of more than 500 killed and 1500 wounded.[118]

Those who actually fought in the campaign have strong opinions about its worth. Peter Medcalf had no doubt that he was fighting a 'useless and unnecessary campaign'. Tod Schacht, of the 9th Battalion, thought that the Australians' escalation of the war on Bougainville was a result of Blamey's desire to achieve some glory during the war's final days. In his history of the 58th/59th Battalion, Russell Mathews, himself a veteran, described the campaign as futile and conjectured that the motivation was more political than military. The historian of the 42nd Battalion labelled the Bougainville battle a 'politician's war'where the men only took risks they could not avoid and 'every life lost was begrudged'. The battalion's history summed up the front line soldier's attitude: 'men fought because there was no alternative. None wanted to lose his life on Bougainville'.[119]

Furthermore, the government's policy of describing the campaign in minimalist terms, in an attempt to hide the actual extent of operations from the Australian public, only served to exacerbate the annoyance of those serving on Bougainville. The combination of infrequent and limited reporting in Australian newspapers, and Curtin's description of the campaign as 'just mopping up', belittled the considerable and dangerous efforts of the fighting troops. It also suggested a continued official preference for highlighting the exploits of the AIF, two divisions of which were fighting in Borneo, to the detriment of the accomplishments of the Militia.[120]

Blamey allocated to Savige the 3rd Division, two independent infantry brigades (the 11th and 23rd), and corps level troops. All his infantry were citizen soldier units. Bridgeford continued as the 3rd Division's commander, and his brigade groups were:

- 15th Brigade with the 24th, 58th/59th and 57th/60th Battalions under the command of Brigadier H.H. Hammer
- 7th Brigade with the 9th, 25th and 61st Battalions under the command of Brigadier J. Field
- 29th Brigade with the 15th, 42nd and 47th Battalions under the command of Brigadier R. F. Monaghan

Of the three brigade groups, only the 15th was a traditional 3rd Division formation; the other two were from Queensland. The 29th Brigade had served briefly under the 3rd Division's command towards the end of the Salamaua Campaign, and the 7th Brigade joined it in Australia during the refit after New Guinea. Of the nine battalions, only the 42nd had served with the 3rd Division during World War I.

Rejoining the 3rd Division in Bougainville were a number of its traditional elements which had been serving under other commands. The 3rd Division Signals sailed from Lae, as did the 15th Field Company which accompanied the 15th Brigade. In addition, elements of the 3rd Division AASC also arrived in Bougainville. The 10th Field Company also made the move from Lae, bringing it once again into close proximity to its Victorian comrades. However, it continued to function as the engineering unit for II Australian

Corps, and did not serve under the 3rd Division's command. Instead, it spent most of its time employed on base construction projects around Torokina.

However, other parts of the 3rd Division continued to serve under other headquarters until the war's end. The 4th Brigade sailed for New Britain, where it again came under the command of the 5th Division and took over from other American units which MacArthur wanted elsewhere. The 4th Brigade served out the war on New Britain where it helped to contain large Japanese forces in Rabaul. In addition, the war's end saw elements of the 3rd Division AASC supporting three different campaigns. Most of the unit was in Bougainville, attached to the 3rd Division, but other components supported the 5th Division on New Britain and the 6th Division on New Guinea. Corporal Thomas D. Hands, for example, served at Aitape and Wewak, on New Guinea, in a transportation platoon.[121]

The Headquarters 3rd Division advance party arrived at Torokina on 6 October. Bridgeford's first brigade to land was the 7th, which disembarked in Empress Augusta Bay in mid-November. Other troops arrived over the next few weeks. By the beginning of February, the Americans had left, except for a few service troops. Savige decided to deploy the 3rd Division on the southern front, against the enemy's main centres of resistance and the island's agricultural districts, upon which the Japanese depended for food. Bridgeford's ultimate objective was Buin, the location of the enemy's headquarters. The 11th Brigade took over the Northern and Central Sector and was ordered to attack towards the Bonis Peninsula, and to push patrols across the Numa Numa Trail to the island's eastern side, to interdict Japanese movements. Initially, the 23rd Brigade received the task of garrisoning the American air and naval installations on the nearby islands of Emirau, Green, Treasury and the New Georgia Group. Eventually, however, Savige moved most of the brigade to Bougainville, where it joined the 11th Brigade on the northern front.[122]

Savige knew he did not have the resources to conduct a major offensive. While the logistical situation had greatly improved on what he had faced at Salamaua, MacArthur's staff refused to allocate more than the minimum amount of sea-lift capacity to an operation so far to the rear. Savige had to plan within his means. In addition, he did not want to waste his men unnecessarily. While he had adopted Blamey's policy of aggressive action, Savige tempered it by insisting that his subordinates keep as few troops as possible in contact with the enemy. On the 3rd Division's sector, this meant that the brigades would fight in turn, rotating between the front, reserve and rest. The Division would keep up constant pressure on the enemy troops, and drive them back towards Buin, but would balance progress with the minimum possible casualties. Savige refused to keep units in the front line which he did not need for combat. Instead, all combat troops were periodically rotated to the rear, out of harm's way, to enjoy a period of proper rest. On Bougainville, therefore, there was no need to repeat the 58th/59th Battalion's marathon 77 consecutive days of contact with the enemy during the Salamaua campaign.[123]

For the veterans of New Guinea, operations on Bougainville were both similar to and different from their previous campaign. Compared to the hardship conditions the troops had endured throughout the struggle in New Guinea, the coming battle represented an easier task. The terrain, for example, while still difficult, was not nearly as extreme. The

24: Troops of 58th/59th Battalion repairing the Buin Road on Bougainville.
[*Courtesy* AWM 094226]

3rd Division would again be operating in a deep tropical jungle, where the dominant factors were heat, humidity and torrential downpours. To Ken Chapman of the 2nd Field Regiment, there was not much difference between the two. Both were hot, wet and muddy.[124] One significant difference, however, was that the towering summits of the Crown Prince Range which dominated the centre of Bougainville served as the campaign's shoulder. This time the men of the 3rd Division would not have to manoeuvre along and over precipitous slopes. Instead, the Division's area of operations was the island's coastal plain and foothills. There, the primary geographic obstacle was the island's many rivers, which raced down from the high ground above the Australians to the sea. Every few miles, one of these rivers, such as the Hongorai or the Mivo, pierced the jungle at right angles to the line of the Australian advance, affording the Japanese a natural barrier behind which to mount a strong defence.

As in New Guinea, the frequent rains on Bougainville ensured that the men fought in a state of perpetual wetness, where uniforms rarely dried and fighting pits became pools of mud. The rains frequently reduced the land to a sodden morass which impeded movement and threatened to swallow those who slipped from the narrow tracks which wound through the undergrowth. Medcalf remembers that, when he was out on one patrol, heavy rains forced his party to sleep in trees, as the ground had disappeared under

25: Troops from the 25th Battalion recover supplies after an air drop on Bougainville. They include: Private E. J. McKee; Private S. B. Kennedy; Private R. V. Hush; Private Miller; Private A. T. Shanahan; Private M. J. Cuskelly; and Corporal R. Strowe. [*Courtesy* AWM 079453]

rising waters. The mud became so extreme that it taxed the engineers' ability to keep the division's tracks and roads serviceable. For example, the engineers corduroyed one stretch of the Buin Road with logs, only to see them sink out of sight into the mud as the rain continued. They then had to cut down more trees and lay a second series of logs. The mud could reach fabulous depths. Harry Finn of the 3rd Division AASC witnessed, one wet day, an enormous bulldozer, sporting six-foot high tracks, plough through the mud near his supply depot. Its tracks were completely submerged, and the mud lapped at the feet of the driver, perched on the top of the vehicle. Behind, he towed a line of 50 supply trailers, each invisible beneath the mud. All Finn could see was 50 ripples, spreading out behind the bulldozer.[125]

On Bougainville, the Australians were able to operate a much improved logistic system which alleviated many of the shortages and hardships the troops had experienced at Salamaua and elsewhere in New Guinea. As the 3rd Division's area of operation was a coastal plain, not mountains, its AASC soldiers were able to bring forward supplies more easily. Although primitive, the island's road system was sufficiently accessible for the operation of trucks and, even more important, jeeps and trailers. Even in forward areas, units generally now received fresh meat once a day. The 3rd Division did continue to

26: Tractor towing supplies through the mud in South Bougainville.
[*Courtesy* Ian Fletcher]

employ some carriers and conduct air drops, but it did not depend as much upon these supply options as it had in New Guinea.

When Savige arrived, he took over a fully operational American base designed to support a force larger than the Australian contingent, and constructed to the usual United States standard of luxury. For example, 3rd Division Signals settled into offices at Torokina that its members could only describe as palatial, compared to their previous experiences. Torokina provided the Australians with a fully equipped and secure base area, including facilities for rest and recreation, enabling units not in the front lines to enjoy a respite from the war. The 3rd Division was even able to establish beachside rest camps, where parties of soldiers went for several days of total rest. Tod Schacht of the 9th Battalion enjoyed a pleasant respite there, relaxing in the dunes under the shade of casuarina trees. The logistic constraints on the campaign were not found on Bougainville, but were imposed by MacArthur's control of shipping, and his determination to direct as many resources as possible to the Philippines.[126]

On 22 November, Savige took responsibility for all operations on the island from Major-General Oscar Griswold, the American commander. Three days later, he announced the Australian change of policy. In the central sector, 50m beyond the 9th Battalion's position on George Hill, was the enemy-occupied Little George Hill. On 25 November,

27: Troops of the 9th Battalion prepare to assault Little George Hill, Bougainville. They are: Private J. W. Uridge; Lance-Corporal H. J. Powell; Private L. W. Callaghan; Corporal H. L. O. Morris; Private F. B. Tuohy; and Private A. M. Hodgson. [*Courtesy* AWM 077305]

the 9th Battalion broke the island's unofficial truce. Under the cover of a smoke bombardment, the Australians launched a surprise attack, secured Little George Hill and killed twenty defenders, at a cost to themselves of two dead and six wounded. In mid-December, the 9th Battalion advanced again, pushing the enemy off the high ground of Artillery Hill. Also in December, the 29th Brigade began the 3rd Division's operations against the southern sector. The 15th and 47th Battalions, supported by the 1st New Guinea Battalion, pushed patrols across the Jaba River and secured the area. Against a startled opponent, unaccustomed to such boldness, the Australians had the upper hand in these actions. The Japanese, however, soon adjusted.[127]

The Australian offensive policy followed a familiar pattern of aggressive patrolling, to reduce the enemy's numbers and morale, to locate his positions and gather intelligence, and to control the jungle. Patrolling laid the basis for the 3rd Division's attacks and enabled it to seize ground, forcing the enemy to retreat. It was not an easy policy to implement, especially for those in close contact with the enemy, despite the official policy of minimum casualties, which meant little to infantrymen slipping out of the relative safety of their company perimeters on a patrol into enemy territory. Moreover, Savige had received American intelligence reports that greatly underestimated the enemy's strength. Instead of 14,000 men, he actually faced 44,000 Japanese soldiers. One report went so far as to suggest that the enemy's soldiers were in a bad state, both physically and materially, and that they would not pose a threat to the Australians.[128]

28: A patrol from 42nd Battalion makes its way through a swamp on Bougainville. [*Courtesy* AWM 078546]

Patrols continued to be tense, lonely missions, carried out by small parties of lightly armed soldiers, who deliberately went behind enemy lines and put themselves in harm's way. Of course, the great size of the area of operations, and the restricted visibility of the jungle, meant that most patrols failed to make contact. The daily operational reports of 42nd Battalion operations show that reports of 'no enemy movement' were the most common outcomes of a patrol. For example, Peter Medcalf was part of an effort to locate and destroy a Japanese camp at Sisiruai. The patrol staggered through a swamp, in water that reached their waists, waging a losing battle against endless leeches and mosquitoes. After three days, they returned to their camp, without having found the enemy's position. When Medcalf removed his boots and socks, the soles of his feet, having been saturated for so long, came off too. When the enemy was found, it was sometimes with little warning. On one occasion, the forward scouts of a 42nd Battalion patrol, under the command of Lieutenant Cole, encountered two enemy soldiers at a range of ten yards. The Japanese scattered as the scouts opened fire. The Australians found some blood stains but the enemy had disappeared. After that, the patrol made no further sightings.[129]

Fortunately on Bougainville, the infantry had greater access than ever before to support weapons. When the 58th/59th Battalion went out on a patrol of the Hatai Track in April 1945, an artillery forward observer accompanied the party. As soon as the patrol located the enemy, the gunner called artillery and mortar fire onto the position. The Australians also had access to tanks and close ground support from four Royal New Zealand Air Force Corsair squadrons, based at Torokina. As a result, the Australians were able to wage a modern, combined arms style of warfare whose firepower the enemy could not match. Moreover, the increased support took some of the burden, and danger, away from the infantry. The 15th Brigade co-ordinated a typical attack on 20 May, advancing towards the Hari River. The 24th Battalion set off down the Buin Road at 8.30 a.m., behind a creeping barrage of artillery, mortar, and machine gun fire and supported by two troops of tanks. The bombardment included more than 1500 rounds of artillery and 700 rounds of mortar fire. Just before the advance, a flight of 80 Corsairs raced in and bombed the enemy's positions. On the 24th Battalion's left flank, the 57th/60th Battalion advanced parallel along the Commando Road, while on the right the 58th/59th Battalion provided support. The offensive took several days, but by its conclusion, Japanese resistance between the Hongorai and Hari Rivers was at an end.[130]

Despite their isolation and hopeless strategic position, the Japanese remained determined opponents, capable of waging a fierce defence and launching vigorous attacks. The Australians might have pushed the enemy back, but at no time did the Japanese concede defeat, and occasionally they even turned and counter-attacked. Early March found the 7th Brigade pushing towards the Puriata. One of its battalions, the 25th, had advanced down the coast by barge and secured the village of Toko before moving inland to continue the advance. On 5 March, it secured a small hill named Slater's Knoll, after a soldier who had been wounded nearby. After a pause, the battalion continued its advance but ran into increasingly heavy Japanese resistance. By 22 March, however, they had driven the enemy from a complex of pillboxes and had opened the way to their opponent's next defensive line, on the Hongorai River. The threat to the Hongorai provoked the enemy. On 27 March the 25th Battalion's forward elements repelled an attack by more than 100 Japanese, while the 9th Battalion fought off other assaults. Other parties of Japanese infiltrated between the Australian positions and cut the Buin Road leading to the rear. On 30 March, the enemy's main attack began and waves of Japanese soldiers attempted to sweep the 25th Battalion off Slater's Knoll. The Australians broke up each attack with concentrated machine gun and tank fire. On 5 April, the Japanese renewed the attack even more fiercely than before. While one party assaulted the knoll's north side, a larger force attempted to overrun the Australians from the south west. Waves of fanatical attackers presented dense targets to the defender's Bren and Vickers gunners, while fire from supporting artillery batteries added to the fury of the defence. No Japanese soldier penetrated the 25th Battalion's perimeter, and only a handful got within a few feet of the Australians' fire pits before being cut down. Japanese bodies lay in heaps around the battalion's defences; the Japanese suffered more than 1600 casualties, including more than 600 dead.[131]

The day after the battle, Ken Chapman, a signaller in the 2nd Field Regiment, had to go to the battlefield to repair the line between the 25th Battalion and the gunners. His unit had fired relentlessly during the Japanese assault and he observed that the terrain around the knoll bore witness to the intensity and accuracy of its fire. It was clear to him that the infantry had been through a terrible ordeal as the men sat around exhausted, unable to talk. He also remembers seeing bulldozers pushing enormous piles of Japanese bodies into mass graves.[132]

In April, the 15th Brigade took the lead in the 3rd Division's drive and the 7th Brigade rotated to the rear for a well deserved rest. By June, the 15th Brigade was approaching the Mivo River, closing in on the Japanese stronghold at Buin. Bridgeford allocated the final drive to the 29th Brigade, which took over from the 15th in early July. On 7 July, a few days after the handover, the Japanese made another determined effort to halt the Australians' advance and to drive them back from the Mivo River. D Company, 15th Battalion bore the brunt of the enemy's fury. William T. Broadhurst was part of a mortar crew which supported the company from within its perimeter. At the height of the battle, the enemy pressed against the Australian wire, threatening to annihilate the position. To bring their guns to bear, the mortarmen removed the secondary charges from their bombs, enabling them to shorten the range and tear holes in the enemy's ranks. The next day, the Australians once again filled mass graves with enemy dead.[133]

As Bridgeford planned for his attack on Buin, nature accomplished what the Japanese had so far been unable to do. On 3 July, the normal daily shower turned into a downpour. A week later, the rain continued unabated and forced Savige to postpone the assault. The 29th Brigade's line of communication became a sea of mud, and Bougainville's many rivers rose and washed away the bridges and fords the engineers had constructed. After another week of solid rain, Savige had again to postpone the attack, this time until the end of August.[134]

Fortunately, the men of the 3rd Division never got the chance to renew the attack. Instead, events at the other end of the Pacific brought operations to a halt. On 6 August 1945, a lone American bomber detonated an atomic bomb over the city of Hiroshima, on the Japanese homeland. Three days later, a second bomb devastated Nagasaki. On 9 August, the Soviet Union declared war on Japan, launching a massive army into Manchuria. Savige was informed of these developments and, on 11 August ordered all patrols to return to their perimeters. Japan officially ceased fire on 15 August. It took Savige several days to convince the enemy that the war was over. It was not until 8 September that a Japanese representative arrived at Torokina, and signed the surrender accord.[135]

* * * *

After six years of world wide conflict, the war was over. The Australian Army now turned to its post-war responsibilities, the internment and repatriation of prisoners of war and the demobilisation of its citizen soldiers. Both processes took some time to complete, because the army had to take charge of tens of thousands of enemy soldiers scattered across a huge area of the South West Pacific. Some fortunate soldiers received their demobilisation orders almost immediately. Peace found the men of the 10th Field Company enjoying a period of leave in Victoria, after more than two years overseas. They had expected to return to Bougainville, but instead, on 11 September, the unit received orders for its disbandment and the men progressively received their discharges. The experience of H. J. Birleson of the 57th/60th Battalion was more typical, however. He was reassigned to the 7th Battalion and transferred to Fauro Island to help guard Japanese prisoners. He then moved to Rabaul and served with the 31st/51st Battalion, returning to Australia only in June 1946. He did not leave the army until November. As it had after World War I, the army disbanded the 3rd Division and its other Militia formations. Their re-raising would have to await the decisions of the government and the army on the organisation of the nation's post-war citizen army.[136]

Throughout their second war of national survival, the citizen soldiers of the 3rd Division had again proved their worth and fulfilled their duty of defending the nation. It had not been an easy task. The years of inter-war neglect had almost destroyed the Militia. Even after war began, the Menzies government only moved slowly to correct the force's deficiencies. After the 2nd AIF's raising, the Militia became the army's step-child. Its neglect continued as the expeditionary force received the best men and equipment. Only after Japan's entry into the war did the government begin to take the improvement of the citizen forces seriously, and by then it was nearly too late. Had it not been for the heroic stand of one of the 3rd Division's own, the 39th Battalion, the Japanese might have reached Port Moresby. Once on the offensive, it was the 3rd Division which defeated the enemy in front of Salamaua, and made the capture of Lae possible. Its men then participated in the drive to clear the Japanese from the Huon region of New Guinea all the way to Madang. On Bougainville, citizen soldiers took the lead in a battle as fierce as anything the AIF had faced in Borneo during the 7th and 9th Division's final campaign. The 3rd Division had begun the war composed of inexperienced, ill-trained and poorly equipped youths, and finished it with highly capable and effective veterans. Their achievements had helped to win victory in the Pacific and safeguard the nation from defeat. The men of the 3rd Division had done their duty.

Endnotes

1 Quoted in Donald Kagan, *On the Origins of War and the Preservation of Peace* (New York: Anchor Books, 1995), p. 405.

2 Information provided by N. W. Hall.

3 Corfield, *Hold Hard Cobbers*, p. 36; and information provided by N. W. Hall.

4 Information provided by Athol Cottrill.

5 '3rd District Command: 'A' Administrative Instruction No. A.11' AWM49, item 148; *Getting Through: The Unit War History of 3rd Australian Division Signals, 1939-1945* (Dandenong, VIC: Kenwalk Printing, 1987), p. 10; Rupert Charlott, ed., *The Unofficial History of the 29th/46th Australian Infantry Battalion AIF, September 1939 – September 1945* (Melbourne, 1952), p. 1; and George Christensen, ed., *That's the Way it Was: The History of the 24th Australian Infantry Battalion (A.I.F.) 1939-1945* (Hawthorn, Vic: 24th Battalion Association, 1982), p. 6.

6 Ron Blair, *A Young Man's War: A History of the 37th/52nd Australian Infantry Battalion in World War Two* (Melbourne: 37th/52nd Australian Infantry Battalion Association, 1992), pp. 17-19.

7 'War Training Instruction No. 1', 19 September 1939, AWM49, item 148.

8 'Instructions for the Training of Militia Units Called out for Special Training', AWM49, item 148.

9 'Responsibilities of the DADOS of a Division', 2 January 1940, AWM49, item 40; 'Shortages of Tents', 21 March 1940, AWM49, item 42; Christensen, *That's the Way it Was*, p. 9; Blair, *A Young Man's War*, p. 20; and information provided by Clarence W. Wood.

10 Christensen, *That's the Way it Was*, pp. 12-13.

11 Information provided by N. H. Taylor, John C. Thomas, A. E. Doyle, George H. Carter and Gordon L. Huggard.

12 Grey, *A Military History of Australia*, pp. 142-43.

13 Blair, *A Young Man's War*, p. 16; and Charlott, *The Unofficial History of the 29th/46th Australian Infantry Battalion AIF*, p. 2.

14 Christensen, *That's the Way it Was*, pp. 8-10; Corfield, *Hold Hard Cobbers*, pp. 36-37; Charlott, *The Unofficial History of the 29th/46th Australian Infantry Battalion AIF*, p. 2; information provided by Saul Spielman; and 'Major General Sir Jack Stevens – Unpublished Manuscript', AWM, 3DRL3561.

15 A. Les Jones, *The Mail Must Get Through* (n.p., 1983), p. 1; and *Getting Through*, p. 12.

16 Information provided by Saul Spielman, Kenneth Ballard, R. F. C. Campbell and Harry Finn.

17 Russell Mathews, *Militia Battalion at War: The History of the 58th/59th Australian Infantry Battalion in World War II* (Sydney: 58th/59th Battalion Association, 1961), pp. 4-5.

18 See F. M. Budden, *The Chocos: The Story of the Militia Infantry Battalions in the South West Pacific Area, 1941-1945* (Summer Hill, NSW: F. M. Budden, 1987), pp. 8-10.

19 Date taken from 'Location Statement', January 1942, AWM52, item 8/2/15.

20 'Letter to Helen', 9 August 1942, Wilton Collection, box 4, folder 7, ADFA.

21 '4th Infantry Brigade War Diary', 8 December 1941, AWM52, item 8/2/4.

22 'Report on Q Activities for the Month of March', and 'Remarks by Transport Officer', 1 April 1942, AWM52, item 8/3/76; and 'Preparedness for War – 15th Brigade', January 1942, AWM52, item 8/2/15.

23 Blair, *A Young Man's War*, p. 55; and information provided by Harry Finn.

24 '24th Battalion War Diary', October 1941-January 1942, AWM52, item 8/3/62.

25 G. M. Keating, 'The Right Man for the Job: An Assessment of Lieutenant General S. G. Savige as a Senior Commander', BA (Hons) thesis, School of History, University College, University of New South Wales, Australian Defence Force Academy (1995), p. 39.

26 'Savige to Hammer', 17 July 1943, AWM54, item 587/7/27.

27 'Officers – 3rd Australian Division, Recommendations for Disposal', December 1942, Savige Collection, 3DRL2529, item 119, AWM.

28 *Ibid.*

29 'Savige to Hammer', 17 July 1943, AWM54, item 587/7/27.

30 'Officers – 3rd Australian Division, Recommendations for Disposal', December 1942, Savige Collection, 3DRL2529, item 119, AWM; 'ADC Diary, December – January 1944', 9 January 1943, Savige Collection, 3DRL2529, item 30, AWM; and 'Letter to Helen', 9 January 1943, Wilton Collection, box 4, folder 7, ADFA.

31 'AMF 3rd Australian Division Administrative Instruction No. 14', 24 June 1942, Savige Collection, 3DRL2529, item 119; and W. B. Russell, *There Goes a Man: The Biography of Sir Stanley G. Savige* (Melbourne: Longmans, 1959), p. 250.

32 Information provided by Harry Wadge.

33 'Preparedness for War – 15th Brigade', January 1942, AWM52, item 8/2/15.
34 Russell, *There Goes a Man*, p. 253.
35 Corfield, *Hold Hard Cobbers*, p. 52.
36 Quoted in Blair, *A Young Man's War*, p. 54.
37 Keating, 'The Right Man for the Job', p. 40, and Christensen, *That's the Way it Was*, p. 16.
38 Blair, *A Young Man's War*, p. 53.
39 Information provided by Gordon L. Huggard and Saul Spielman.
40 Corfield, *Hold Hard, Cobbers*, pp. 57-58.
41 Information provided by Rayden T. Tudor, Allen G. Martin and Robert I. Caporn.
42 'Appreciation of the Situation', 26 May 1942, AWM54, item 243/6/99.
43 *Ibid.*
44 *Ibid.*
45 Corfield, *Hold Hard, Cobbers*, p. 58; Mathews, *Militia Battalion at War*, p. 9; Charlott, *The Unofficial History of the 29th/46th Australian Infantry Battalion AIF*, p. 5; and '37th Australian Infantry Battalion Administrative Instruction', AWM52, item 8/3/76.
46 'Officers – 3rd Australian Division, Recommendations for Disposal', December 1942, Savige Collection, 3DRL2529, item 119, AWM; and Blair, *A Young Man's War*, pp. 60-63.
47 Information provided by Douglas A. Wells.
48 '37th/52nd Battalion War Diary', September 1942, AWM52, item 8/3/76; Jack Bourne and Jack Lavery, *A Purple Patch: A History of the 10th Field Company, Royal Australian Engineers (AIF) 1940-1945* (Melbourne: 10th Field Company, 1991), p. 34; and Alfred N. Festburg, *The Lineage of the Australian Army* (Melbourne, Allara Publishing Pty. Ltd., 1972), pp. 96, 110, 114-15.
49 'C. O'.s Review on Disbandment of 59th Australian Infantry Battalion', 18 September 1942, AWM52, item 8/3/95; '37th/52nd War Diary', September 1942, AWM52, item 8/3/76; and information provided by Jeffrey Pooler, Royce G. Ellis, David G. Taylor, C. H. Howe. See also, Mathews, *Militia Battalion at War*, pp. 8-9; and Blair, *A Young Man's War*, p. 70.
50 The 39th Battalion's history is told in Victor Austin, *To Kokoda and Beyond: The Story of the 39th Battalion, 1941-1943* (Carlton, VIC: Melbourne University Press, 1988).
51 '39th Battalion War Diary', AWM52, item 8/3/78; '15th Brigade War Diary', 23 January 1942, AWM52, item 8/2/15; and Austin, *To Kokoda and Beyond*, pp. 9-10.
52 '39th Battalion War Diary, AWM52, item 8/3/78.
53 Diary, Mowat Collection, 3DRL7137, AWM.
54 Quoted in Dudley McCarthy, *South-West Pacific Area – First Year, Kokoda to Wau* (Canberra: Australian War Memorial, 1962), p. 207.
55 *Ibid.*
56 '39th Battalion War Diary', AWM52, item 8/3/78.
57 Charlott, *The Unofficial History of the 29th/46th Australian Infantry Battalion, AIF*, p. 8; Bourne and Lavery, *A Purple Patch*, pp. 64-66; and information provided by Lawrence Armstrong.
58 '58th/59th Battalion War Diary, March – April 1943', AWM52, item 8/3/95; and information provided by John C. Thomas and Harry Finn.
59 'Report on New Guinea Operations', Moten Collection, PR91/002, item 8i, AWM; and 'Precis of a Report on the Construction of the Bulldog-Wau Road and Lessons Learned', AWM54, item 863/2/12.
60 '58th/59th Battalion War Diary, March – April 1943', AWM52, item 8/3/95; information provided by Geoff Waters; Russell, *There Goes a Man*, pp. 266-67; and Christensen, *That's the Way it Was*, p. 32.
61 Information provided by Garry A. McGuire.
62 'Report on Operations of 3rd Australian Division in Salamaua Area From 22 April 43 to 25 August 43', Blamey Collection, 3DRL6643, item 3/14, AWM. *See also*, Russell, *There Goes a Man*, pp. 259-60.
63 A. R. Ross, ed., *The Magazine of the 17th Australian Infantry Brigade, 1939-1944* (n.p. 1944), p. 38.
64 'Report on Operations 3rd Aust. Div. – Salamaua Area, 22 April – 25 August 1943: Appreciation, Orders, Instructions, Vol. 1', Savige Collection, 3DRL2529, item 51, AWM; and 'Report on New Guinea Operations', Moten Collection, PR91/002, item 8i, AWM. *See also*, Russell, *There Goes a*

Man, p. 270; and David Dexter, *The New Guinea Offensives* (Canberra: Australian War Memorial, 1961), pp. 20-21, 58.

65 Quoted in Russell, *There Goes a Man*, p. 262.

66 Information provided by John C. Thomas.

67 Information provided by N. H. Taylor, Brian J. Balcombe, Allen G. Martin, William Mateer, Saul Spielman and Clarence T. Tonkin.

68 Information provided by Jeffrey Pooler, Russell Mathews, John J. Grimes, and Saul Spielman.

69 Information provided by Saul Spielman.

70 S. E. Benson, *The Story of the 42nd Australian Infantry Battalion* (Sydney: 42nd Australian Infantry Battalion Association, 1952), p. 54; and information provided by Allen. G. Grantly, Robert Ian Caporn, Kenneth Ballard and John C. Thomas.

71 Information provided by Jeffrey Pooler, John J. Grimes, David G. Taylor, E. G. Limbrick, Thomas J. Cahill, Colin F. Lilley, Graeme T. Mills, C. H. Howe and Harry Finn. *See also*, 'Greenlough to Aunt and Uncle', 28 October 1943, Greenlough Collection, 3DRL6461, AWM; and Mathews, *Militia Battalion at War*, pp. 15-24.

72 Frank Taylor, *Mad Mick's Mob: A History of the 15th Australian Field Company, RAE (AIF)* (Mordialloc, VIC: 1995), pp. 47-53.

73 'Report on New Guinea Operations', Moten Collection, PR91/002, item 8i, AWM.

74 Information provided by C. R. Meilroy.

75 'Report by Lieutenant A. W. Bunsell', 5 April 1943, AWM52, item 8/3/62; and information provided by David Tomlin and Harry Finn. *See also*, 'Report on New Guinea Operations', Moten Collection, PR91/002, item 8i, AWM.

76 'Staff Directive – Planning Future Operations in Wau Area', 15 February 1943, AWM54, item 583/3/3; 'Report on Establishment of Moresby-Bulldog-Wau Line of Communications', 1 April 1943, AWM54, item 583/7/3; and information provided by Clarence Tonkin and Colin F. Lilley.

77 Information provided by Kenneth Ogier and Harry Finn.

78 Information provided by Harry Finn.

79 Information provided by E. G. Limbrick.

80 'The Scope of Reciprocal Lend-Lease', 5 December 1942, AA(ACT), A5954/69, item 2386/3.

81 Information provided by Douglas Wells.

82 'Baker to Home', 16 July 1943, Baker Collection, 3DRL7528, AWM; and Taylor, *Mad Mick's Mob*, p. 51.

83 Information provided by Alan Hulme and John C. Thomas.

84 'Report on Operations 3rd Australian Division, Salamaua Area, 22 April–25 August, 1943: Administration, Supply, Medical', Savige Collection, 3DRL2529, item 60, AWM; 'Report on New Guinea Operations', Moten Collection, PR91/002, item 8i, AWM; 'Health of Forward Troops in Kanga Force', 4 April 1943, AWM52, item 25/3/5; and information provided by Douglas Wells.

85 'Report on New Guinea Operations', Moten Collection, PR91/002, item 8i, AWM.

86 Information provided by Allen G. Martin.

87 Information provided by John. C. Thomas.

88 Information provided by Geoff Waters.

89 'Signals in Salamaua Area – New Guinea', 19 October 1943, AWM52, item 7/16/3; and information provided by E. G. Limbrick, Douglas A. Wells, Robert I. Caporn, and John J. Grimes.

90 Dexter, *The New Guinea Offensives*, pp. 25-31; and Russell, *There Goes a Man*, p. 260. *See also*, 'Report on Operations of the 3rd Australian Division in Salamaua Area – Appendix D, Policy for Operations in Mubo Area', (author's possession).

91 'Report on Operations of the 3rd Australian Division in Salamaua Area – Appendix D, Policy for Operations in Mubo Area' (author's possession).

92 'Commanding Officer's Report for Month of May 1943 – 24th Battalion', AWM52, item 8/3/62. *See also*, S. G. Savige, *Tactical and Administrative Doctrine for Jungle Warfare* (HQ 2 Aust Corps, 1945), pp. 8-10.

93 Information provided by Royce G. Ellis and Jeffrey Pooler.

94 'Commanding Officer's Report for the Month of August', 31 August 1943, AWM52, item 8/3/62.

95 'Report on the Lower Watut Patrol, 7 May–26 May 1943', AWM52, item 8/3/62; and Christensen, *That's the Way it Was*, pp. 52-4.
96 Mathews, *A Militia Battalion at War*, p. 66.
97 Information provided by William Mateer.
98 Information provided by Harold Hibbert; and 'Report of Activities of Standing Patrol – OP on New Asini Village', 10 August 1943, AWM52, item 8/3/62.
99 Unless otherwise noted, this section is taken from 'Report on Operations of the 3rd Australian Division in Salamaua Area', (author's possession)
100 '58th/59th Battalion War Diary', AWM52, item 8/3/95; and Mathews, *Milita Battalion at War*, pp. 25-43.
101 Benson, *The Story of the 42 Australian Infantry Battalion*, pp. 63-66.
102 For an assessment of Savige's relief, *see* Keating, 'The Right Man for the Job', pp. 62-71.
103 Russell, *There Goes a Man*, pp. 281-85.
104 *Getting Through*, pp. 36-38, 43-45.
105 Mathews, *Militia Battalion at War*, p. 99; and Jones, *The Mail Must Get Through*, p. 20.
106 For a summary of these events see, 'Report on Operations – Ramu Valley', and 'Report on Operations – Mindjim Valley, Kabenau River, Bogadjim and Madang', Savige Collection, 3DRL2529, item 46, AWM.
107 For a summary of these events, *see* Blair, *A Young Man's War*, pp. 157-252 and; Charlott, *The Unofficial History of the 29th/46th Australian Infantry Battalion*, pp. 31-108.
108 Information provided by Alan Hulme and Harry Finn.
109 Benson, *The Story of the 42nd Australian Infantry Battalion*, pp. 142-43, 146-47; and information provided by Alan Hulme.
110 Quoted in Christensen, *That's the Way it Was*, p. 172.
111 Grey, *A Military History of Australia*, p. 184.
112 For some examples, *see* Peter Charlton, *The Unnecessary War: Island Campaigns of the South-West Pacific, 1944–45* (South Melbourne: MacMillan, 1983); Harry A. Gailey, *Bougainville 1943-1945: The Forgotten Campaign* (Lexington, KY: University Press of Kentucky, 1991), pp. 192-93; Grey, *A Military History of Australia*, pp. 184-86; and Joan Beaumont, 'Australia's War: Asia and the Pacific' in Joan Beaumont, ed., *Australia's War, 1939-45* (St. Leonards: Allen and Unwin, 1996), pp. 46-47.
113 David Horner, *Blamey: The Commander in Chief* (St Leonards: Allen & Unwin, 1998), pp. 520-2; and John Hetherington, *Blamey: The Biography of Field-Marshal Sir Thomas Blamey* (Melbourne: F. W. Chesire, 1954), pp. 220-21.
114 Charlton, *The Unnecessary War*, pp. 1-2.
115 Hetherington, *Blamey*, p. 217.
116 John Robertson, *Australia at War: 1939-1945* (Melbourne: William Heinemann, 1981), pp. 172-73.
117 Hetherington, *Blamey*, p. 217.
118 'Statement Presented by the Right Hon. F. M. Forde, Deputy Prime Minister and Minister for the Army, regarding Australian War Effort', Wilton Collection, box 4, folder 26, ADFA.
119 Peter Medcalf, *War in the Shadows: Bougainville, 1944-45* (Sydney and Canberra: Collins and Australian War Memorial, 1986), p. 87; and Todd Schacht, *My War on Bougainville: War Under the Southern Cross* (Loftus: Australian Military History Publications, 1999), pp. 133-34; Mathews, *Militia Battalion at War*, p. 147; and Benson, *The Story of the 42nd Aust. Inf. Bn.*, p. 157.
120 Medcalf, *War in the Shadows*, pp. 86-87.
121 Bourne and Lavery, *A Purple Patch*, pp. 93-104; *Getting Through*, pp. 48-49; and information provided by Thomas D. Hands.
122 Russell, *There Goes a Man*, p. 291; and Gailey, *Bougainville*, pp. 193-94.
123 Russell, *There Goes a Man*, pp. 287-88.
124 Information provided by Ken Chapman.
125 Medcalf, *War in the Shadows*, pp. 26-27; and information provided by Harry R. Finn.
126 *Getting Through*, p. 50; Mathews, *Militia Battalion at War*, pp. 166-67; and Schacht, *My War on Bougainville*, p. 182.
127 Gailey, *Bougainville*, pp. 195-96; and Gavin Long, *The Final Campaigns* (Canberra: Australian War Memorial, 1963), pp. 108-12.

128 Russell, *There Goes a Man*, p. 289; and 'Appreciation of the Situation', 20 July 1944, Savige Collection, 3DRL2529, item 48, AWM.

129 'Cyclone and Tempest Bulletins', Byrne Collection, PR00446, AWM; and Medcalf, *War in the Shadows*, pp. 26-27.

130 Long, *The Final Campaigns*, pp. 190-91; Mathews, *Militia Battalion at War*, p. 164; and '24th Australian Infantry Battalion Report on Operations: South Bougainville, January 1945 to July 1945' (author's possession).

131 Gailey, *Bougainville*, pp. 202-4.

132 Information provided by Ken Chapman.

133 William Broadhurst, *The Young Soldiers: The Battle of the Mivo Crossing, the 1945 Experience of the 15th Battalion Mortar Platoon* (n.p., n.d.), pp. 2-4.

134 Gailey, *Bougainville*, pp. 207-9.

135 'Report on the Surrender of Japanese Forces on Bougainville and Adjacent Islands', Blamey Collection, 3DRL6643, item 2/23.91, part 1, AWM.

136 Bourne and Lavery, *A Purple Patch*, pp. 106-7; and information provided by H. J. Birleson.

Chapter 5
The 3rd Australian Division During the Cold War

Post-War Reorganisation of the Australian Army

Victory over Germany and Japan in World War II did not bring security or the disappearance of military responsibilities. Instead, the peace ushered in a period of defence uncertainty and ideological rivalry between the superpowers, the United States and the Soviet Union, and their allies and client states. As a consequence Australia faced sizeable ongoing military commitments, that had not been the position after the World War I settlement.

These new defence commitments included immediate tasks, such as the care and repatriation of tens of thousands of enemy prisoners of war, the restoration of local and colonial governments, and the recovery and storage or sale of mountains of surplus equipment. In the mid-term, the government also took a leading role in the Empire's contribution to the occupation of Japan. Australia took over command of what became known as the British Commonwealth Occupation Force, and contributed a brigade group to the imperial contingent. At its peak, about 10,000 Australians were serving in Japan. Last, arising from its post-war commitments, and made more urgent by the need to help deter or repel the spread of communism, the government also accepted the necessity of having a body of permanent combat troops which it could deploy overseas. The acceptance of a standing force proved a major and long-term change in national defence policy.

Another major post-war change in Australia's security policy was the government's overdue recognition that Britain was unable to provide for the Commonwealth's defence. The failure of the Singapore Strategy underscored Britain's weakness and the decline of its military power. While the British still had a role to play in South East Asia, especially in Malaya, the government of Prime Minister Ben Chifley knew that it had to make other arrangements to secure the nation's defence.

For these reasons Chifley, and his successor Robert Menzies, followed a two-pronged defence initiative. Their first concern was to participate in international arrangements, such as those sponsored by the United Nations and others, including the SEATO and ANZAM agreements for the defence of South East Asia and Malaya, respectively. The government hoped that, by joining other nations in regional pacts, it might provide for the nation's defence on a collective basis. The second policy was to formalise a long-term defence relationship with the United States, as it was obvious after the war's end that

America was the only nation capable of providing for Australia's defence against a major opponent. In 1951, these efforts bore results with the signing of the ANZUS Treaty between Australia, New Zealand and the United States, an agreement which remains the cornerstone of Australia's security arrangements.

As it had done in 1919 the government quickly discarded the army's existing organisation after the surrender of Japan in 1945, pending a decision on the nation's new security obligations and the outline of a new structure for the military. The army disbanded its militia formations, including the 3rd Division, as well as those of the 2nd AIF. As Australia still had considerable military personnel serving overseas, the government placed all its various forces into a temporary institution, which it called the Interim Army. The responsibility of the Interim Army was to manage current military tasks, including demobilisation, while laying the basis for a new permanent army organisation. The Interim Army was always seen as a short term expedient, to serve as a transitional military organisation until the government authorised a permanent post-war structure.

On 3 June 1947 the government gave its approval to the army's new organisation. The force structure consisted of:

- a permanent field force of one independent brigade group of three infantry battalions, plus supporting armoured, artillery, engineer, signals and logistic units
- a citizen soldier field force of two infantry divisions, an independent brigade group, an armoured brigade and corps units
- the necessary fixed establishments to command and administer the force including troops for the fixed defences, base, and training establishments.[1]

At that time the official name for Australia's citizen forces, formerly the militia, became the Citizen Military Forces or CMF. In September 1947, the Minister for the Army, Cyril Chambers, also authorised a name change for the PMF to the Regular Army. Shortly thereafter, he further modified the name, to the Australian Regular Army or ARA. Under the plan, the army's strength was 69,000, made up of an ARA of 19,000 and a CMF of 50,000.

The army's leaders had initially asked the government for a considerably larger force than they ultimately received. In early 1946, the Vice Chief of the General Staff, Lieutenant-General Sydney Rowell, had requested a force of five divisions, including enough permanent soldiers to form one division. It was not an unreasonable request, given the insecurity of the post-war environment and the force's already considerable commitments.

Rowell's proposal was similar in scale to the inter-war army, and considerably smaller than the force the nation had maintained during World War II. Yet his proposal failed, largely on fiscal grounds. The Council of Defence had estimated that Rowell's scheme, in addition to the requirements of the Royal Australian Navy and Royal Australian Air Force, required a defence vote of about £90 million. Chifley recoiled from this amount and instead arbitrarily set the defence estimate at £50 million. From that, the Minister for

Defence, John Dedman, had to fund the three services, provide for his portfolios and the three service ministries' administrative expenses, and support a host of research and technology committees. For the army's leaders, the force structure Chifley subsequently approved was the best organisation affordable with the funds provided. As so often in the past, the government put Treasury preferences ahead of defence concerns, and the army organisation was a product of fiscal considerations rather than military requirements.

The government also denied several of the Military Board's other key requests. The army had asked for the resumption of conscription but instead would have to raise its force through recruitment. More seriously, it failed to secure changes to the *Defence Act*, in particular modification of the clause prohibiting overseas service by the CMF without each soldier's individual permission. Despite the inefficiency of this practice, so clearly demonstrated in two great conflicts, the army, alone among the services, still had to seek the permission of citizen soldiers before they could serve outside Australia. As a result, all troops participating in the occupation of Japan, and later Korea, had first to specifically attest their willingness to go.

From the perspective of citizen soldiers the most significant change in the army's post-war organisation was the raising of a standing field force of regular soldiers. In the past, this had been a responsibility that the *Defence Act* reserved to the CMF, and therefore posed a legal problem for the government. One small precedent existed, however, to get around the letter of the law. On the eve of World War II, the Menzies government had legalised the creation of the Darwin Mobile Force by enlisting its members as gunners or engineers, rather than the prohibited infantrymen or cavalrymen. This time, to get around the *Defence Act*, Chifley employed a different expedient. The Governor-General had not yet revoked his 'Time of War' proclamation, and Australia was technically still in a state of war, thereby obviating the prohibition. Not until 1950 did Menzies change the *Defence Act*, and place the permanent field force on a secure legal basis.

The creation of a permanent field force had tremendous implications for the organisation of the army, the relationship between the CMF and ARA, and the role of the 3rd Division. Australia's commitments to help its allies, and to deter communist aggression more broadly, mandated that the army maintain a force capable of rapid deployment overseas. This was a task which only soldiers undergoing full-time training could perform with the required speed and effectiveness. As the only branch of the army capable of this, the regulars had a relatively easy means by which to define a mission and claim a stake in national security policy.

By contrast, the CMF had far more difficulty identifying a post-war role in the nation's defence. At various times, its purpose had been to serve as an expansion base for a major war, a training establishment for leaders, or a reserve of semi-trained soldiers for the ARA. More often than not, however, no policy existed at all on which to base the CMF's formations, the unit's training programs and the development of their members military capabilities. Major-General Kevin Cooke, a former commander of the 3rd Division and a Chief of Army Reserve, has pointed out that one of the gravest problems the CMF faced after World War II was the failure of the army and the government to articulate or support a clearly defined and consistent role for its citizen forces.[2]

The lack of a clear task for the CMF in Australia's defence left citizen soldiers at a grave disadvantage. Over time, the ARA was able slowly to supplant the citizen soldiers as the primary branch of the Australian Army. Moreover, regular officers came to dominate policy making and the allocation of resources within the force. The results of these developments were devastating for the CMF, and for the 3rd Division.

The 3rd Division is Reborn

On 1 July 1948, recruitment began for the CMF. The army designated its divisions as the 2nd, based on New South Wales, and the 3rd, centred on Victoria. The army raised the independent infantry brigade, the 13th, in Western Australia, and located the headquarters for the armoured brigade at Puckapunyal. After an absence of less than three years, the 3rd Division was once again on the army's order of battle.

As the CMF now fielded less than the equivalent of three divisions, rather than the seven it had possessed in 1920, the 3rd Division had to widen its peace-time catchment area well beyond the traditional heartland, Victoria, to include South Australia and Tasmania. For much of its remaining history, the 3rd Division maintained an association with these less populated States.

Despite the opportunity to build on the citizen soldier reputations established over the course of World War II, Army Headquarters once again assigned units to the 3rd Division on the basis of geographic convenience rather than history. Part of the problem was that the post-war army had far fewer battalions than during the war or in 1920. Therefore, most battalions disappeared from the reduced order of battle. However, despite this limitation, the army still played little heed to traditional associations. For example, the 9th, 25th and 42nd Battalions, which had fought under the 3rd Division's command in World War II, were allocated to the 2nd Division, since they were Queensland-based regiments. Of the three battalions, the 42nd was unique, as it was the only unit which had fought under the 3rd Division in both world wars. The army also failed to re-raise the 39th Battalion, despite its heroic stand on the Kokoda Track. As now constituted, only the 58th Battalion, as part of the linked 58th/59th Battalions, had belonged to the 3rd Division during its New Guinea and Bougainville campaigns in World War II. Only the 40th Battalion from Tasmania, in the linked 12th/40th, and the 38th Battalion from northern Victoria had fought with the 3rd Division in World War I. The new 3rd Division now bore little resemblance to its previous wartime organisation.[3] Table 12 outlines the organisation of the 3rd Division in 1948.

Recruitment into the CMF was slow and, after a year, the force had attracted only 14,000 soldiers out of an establishment of 50,000. By 1951 the figure was still below 21,000 or only 42 per cent of objective. In part, this reflected a population weary of war after six years of conflict. However, government policies did not make military service

Table 12
Order of Battle – 3rd Division, 1948[4]

TYPE	COMPONENT
Headquarters	3rd Division Headquarters
Armoured	3rd Reconnaissance Regiment
Artillery	Headquarters, 3rd Division RAA
	2nd Field Regiment
	10th Field Regiment
	13th Field Regiment
	6th Anti-Tank Regiment
	2nd Light Anti-Aircraft Regiment
	3rd Division Locating Battery
Engineers	3rd Field Engineering Regiment
Signals	3rd Division Signal Regiment
Infantry	4th Brigade Headquarters
	5th Bn (The Victorian Scottish Regiment)
	6th Bn (The Royal Melbourne Regiment)
	8th/7th Bn (The North Western Victorian Regiment)
	6th Brigade Headquarters
	38th Bn (The Northern Victorian Regiment)
	58th/32nd Bn (The Melbourne Rifles)
	12th/40th Bn (The Tasmanian Regiment)
	9th Brigade Headquarters
	10th Bn (The Adelaide Rifles)
	27th Bn (The South Australian Scottish Regiment)
Supply & Transport	Headquarters 3rd Division Column
	6th, 7th, 8th, 9th Transport Companies
	3rd Division Headquarters Transport Platoon
Medical	3rd, 4th, 6th Field Ambulances
Ordnance	Headquarters 3rd Division AAOC
	3rd Division Ordnance Field Park
	3rd Division Mobile Laundry and Bath Unit
Electrical & Mechanical	Headquarters 3rd Division AEME
	105th, 106th, 107th Infantry Workshops
	3rd, 4th, 12th, 13th, 14th, 22nd, 23rd, 24th, 29th, 31st LADs
	3rd LAA Workshop
Postal	3rd Division Postal Unit
Provost	3rd Division Provost Company

Note: Subsequently, the army reraised the 59th Battalion (The Hume Regiment) and allocated it to the 6th Brigade, moving the 12th/40th Battalion to the 9th Brigade.

terribly attractive. For example, the government at first refused to grant public servants leave with pay to attend training camps, and citizen soldiers had to pay their own public transport fares to attend parades. In addition, Chifley insisted on subjecting citizen soldier pay to income tax, and refused to grant an efficiency bonus to reward the force's most dedicated soldiers.

Members of the CMF knew that the government's policies were self-defeating. Lieutenant-Colonel Colin H. Brown, a veteran of World War II and Korea, who spent the last years of his military career in the 58th/32nd Battalion, pointed out that the tax on pay only served to push citizen soldiers into a higher bracket, making recruitment and retention more difficult. Chambers, the Minister for the Army, agreed with this assessment and pointed out to the government that its insistence on taxing a citizen soldier's pay hindered the army's efforts to attract recruits. His pleas, however, failed to move the Treasury. Eventually, the government did make some concessions, reversing its policy on public servant leave and the reimbursement of fares, but its overall objective was still one of enforced frugality – Treasury's traditional response to military funding during peacetime. The government did not lift the tax on citizen soldier pay until 1964.[5]

Although 3rd Division recruitment was poor, the formation did attract a core of World War II veterans. These highly qualified soldiers provided the Division with leaders experienced in first-hand operations and combat – a great advantage when training raw recruits. Their presence allowed units to achieve more quickly a level of effectiveness which would have normally taken far longer to develop. Among them was Robert N. Thompson, who began his military career in the 57th/60th Battalion before volunteering in 1940 for the AIF. He then served throughout World War II with the 2nd/14th Battalion in the Middle East, New Guinea and Borneo, reaching the rank of company commander. After the war, he joined the 58th/32nd Battalion as its second in command before serving on the 3rd Division staff. Similarly, Alex Lochhead enlisted in the 39th Battalion and served in its ranks till its disbandment in July 1943. He then continued his service in the 2nd/2nd Battalion. After the war, he enlisted in the 58th/32nd Battalion. R. F. C. Campbell spent most of the war as a signaller posted to the 1st Armoured Division and, upon its disbandment, moved to 3rd Division Signals, serving in the Bougainville campaign. When recruitment began in 1948, he again came forward and joined the 3rd Division Signal Regiment with the rank of captain.[6]

Veterans also served in the 3rd Division's senior ranks. I. H. Lowen had joined the militia as a private soldier in 1935 and served in the 5th Battalion. At the outbreak of World War II, he volunteered for the AIF and saw action in the Middle East and across the South West Pacific Area, in various regimental and staff appointments. When the army re-created the CMF, he returned to the 5th Battalion as a major, and became its commander in 1950. Lowen reached the rank of brigadier and commanded the 3rd Division's 4th Brigade.[7]

Another senior officer who helped to re-establish the 3rd Division was Lieutenant-Colonel R. R. Gordon. During World War II, he had served with the 2nd/3rd Machine

Gun Battalion, eventually becoming its commander. In 1948, he joined the CMF, receiving the command of the 6th Battalion. Reflecting on the battalion's raising, he wrote that he had had no trouble attracting a pool of experienced officers. The unit reached a strength of 680, nearly its full establishment – a rare event for the CMF in the post-war period. Gordon pointed out that his battalion was the strongest in the army. By the end of his career, he reached the rank of major-general and commanded the 3rd Division from 1960–63.[8]

Recruitment problems did not affect only the CMF, but also the ARA. During the Korean War, Australia could not have provided any more than a token contingent to the United States-led coalition. In fact, at the conflict's start, the manpower level among the regulars was so poor that to bring 3 RAR, the unit chosen for Korea, up to strength, the ARA had to cannibalise the rest of its order of battle. Periodically, during the war, the army had to resort to expedients such as special recruitment drives specifically targeting World War II veterans. The low level of Australia's contribution to Korea was in marked contrast to the vast force the nation had supported during World War II.[9]

The experience of a few veterans, however, was not enough to compensate for the low strength of the CMF units. Had the Korean War escalated, mobilisation of the CMF would soon have revealed the dire situation. The shortage of manpower was of some concern to the government, but it was not until after the outbreak of the conflict in Korea that it received political attention.

In 1951, to solve the army's manpower problems, the Menzies government proposed the reintroduction of conscription. More important, it accepted the need for a lengthy training period if the scheme were to produce effective soldiers. The soldier's obligation was now an initial 98 days of continuous full-time duty in a National Service Training Battalion, followed by 176 days training with a CMF unit over three years. Upon completion of CMF service, the soldier passed into the reserves for a further five years. Voluntary enlistment in the CMF was also an option.

The principal objective was to bring the CMF units up to full strength. For the ARA, however, the conscription scheme was a burden. The regulars had to devote much of their limited resources to training conscripts who then spent most of their military careers in the CMF. For regulars, national service was a liability, and instead of improving its combat capability, the effectiveness of ARA units actually declined. In effect during this period the ARA had to serve two masters. First, it was the trainer of national servicemen and indirectly of the CMF. Second, it provided Australia's contribution to Cold War tasks, and this in a period when its resources were inadequate for even one of these jobs.

For the CMF, national service was a great boon, and the effect on citizen forces units was dramatic. The CMF's ranks quickly swelled and by 1953, the force had a strength of over 80,000 men. As a result, the CMF could man its units at full strength, and was able to raise two more infantry brigades and a host of corps-level artillery units. Moreover, the units had a full allocation of equipment, as the army could draw upon its vast stock of World War II surplus materials. To be fully equipped and manned was indeed unusual for citizen forces in peacetime, and the 1950s were their heyday.

Two highlights stand out in the 3rd Division's training during this period. The first was a divisional parade organised at Puckapunyal on 23 March 1958 by Major-General H. H. Hammer, the 3rd Division's commander from 1956–59. Hammer was a veteran of the 3rd Division, having served as the commander of 15th Brigade in New Guinea and Bougainville during World War II. He believed that it was important for the CMF to be visible and to demonstrate its strength and capabilities. The parade brought together all the Division's Victorian units, namely the 4th and 6th Infantry Brigades, 2nd and 15th Field Regiments, 3rd Field Engineer Regiment, 3rd Division Signal Regiment and elements of the formation's other corps. This was an unprecedented concentration of men and material, recalling the formation's inspection by the King in 1916 on Salisbury Plain. Adding gravity to the occasion was the presence of the Chief of the General Staff, Lieutenant-General Henry Wells, as reviewing officer, as well as a stand filled with dignitaries.[10]

Building on the experience of the parade, Hammer then set for the 3rd Division's show piece the next year the most ambitious and important exercise that the citizen forces had ever held during peacetime. On 14-15 March 1959 at Puckapunyal, the 3rd Division undertook a brigade group exercise centred on the 4th Brigade. Hammer called it Operation Firepower. The exercise involved more than 3500 men, 510 vehicles, thirteen tanks, 24 field guns and other supporting arms. Hammer also invited some 1500 visitors. They included ministers, members of parliament, senior officers of the three services, and leaders of Australian industry and commerce. Therefore, in addition to being a training exercise, Operation Firepower was an opportunity for the CMF to demonstrate its capabilities to a wide and influential audience. Table 13 lists the 3rd Division units which took part.

The General Officer Commanding Southern Command, Lieutenant-General H. G. Edgar, explained that the purpose of the exercise was to design and demonstrate a brigade-level battle. In his comments he noted just how rare an opportunity this was for the CMF. The exercise featured a covering force action, a battalion withdrawal and a counter-attack, among other activities. Units were at full strength, which allowed officers to exercise command under appropriate conditions. Hammer also used the event to drive home the point that firepower dominates the battlefield, and units received authorisation to discharge munitions on a scale unprecedented in peace time. He allowed the gunners, in particular, free reign in the firing of shells (a decision helped by the fact that surplus artillery ammunition was nearing the end of its safe storage life).[11]

Operation Firepower proved a great success. Lieutenant Jim Killender participated, with the 2nd Field Regiment, and he remembers the regiment's guns firing at a far greater rate, and for a longer period, than they ever had in the past. In his memoirs, he drolly noted that in his command post was a large container of aspirins to which the staff helped themselves as needed. Peter Charlton, an eighteen-year old who had just finished his national service training, commented that it felt good to learn something like this instead of drill, and that it was a worthwhile event. For Alan Walsh, another young soldier, Operation Firepower was a chance to see all the arms of the army in operation and how they fitted together in combat. He also appreciated the widespread use of live ammunition and the excitement it added to the exercise. To the 4th Brigade's commander, Brigadier I. H. Lowen, Operation Firepower showed what could be done by citizen soldiers when they were given proper training, equipment and leadership.[12]

Table 13
Units Participating in Operation Firepower[13]

TYPE	COMPONENT
Headquarters	Headquarters 4th Brigade
Armour	A Squadron 4th/19th The Prince of Wales's Light Horse
Artillery	2nd Field Regiment
	Q Battery, 10th Medium Regiment
Engineers	38th Field Squadron
Signals	E & J Troops, 3rd Division Signal Regiment
	One Tentacle, 1st Air Support Signal Unit
Infantry	5th, 6th, 58th/32nd Battalions
Supply & Transport	6th Transport Company, RAASC
Medical	4th Field Ambulance
Ordnance	7th Ordnance Field Park
Electrical & Mechanical	106th Infantry Workshop Stores Section
	106th Infantry Workshop
	2nd Field Regiment LAD
	4th Infantry Brigade LAD
	4th/19th The Prince of Wales's Light Horse LAD (det)
	6th Transport Company LAD
Provost	2 sections, 3rd Division Provost Company
RAAF	21st City of Melbourne Fighter Squadron
	Section of 1st Aviation Company

Despite Operation Firepower's accomplishments, coming changes in national security direction ensured that the 3rd Division would never again have the ability to organise, or even pay for, an exercise on such a scale. Unfortunately for the 3rd Division, the heyday of the CMF had been all too brief. The first of numerous adverse policy changes occurred in 1957, changes which gradually eroded the capabilities and status of the CMF.

In that year, at the request of the CGS, the government modified the national service scheme, reducing its intake from 29,250 to just 12,000 trainees per year. Moreover, trainees would now receive only 77 days of training from ARA instructors in National Service Training Battalions, before passing into the CMF. The objective of this reform was to raise 1st Brigade for the ARA, not to help the CMF. It relieved the regulars of much of their instructional responsibilities, and freed up personnel the ARA could use elsewhere. The ARA's field force was greatly under-strength and, by reducing its national service obligation, the army and the government hoped to improve its combat readiness and capability.

At the same time as it modified the national service scheme, the government agreed to reduce the CMF's establishment to just 51,000 soldiers from an existing enrolment of approximately 80,000. As a result, the army once again had to set for its CMF units posted strengths far below their wartime establishments. Another concern for the CMF was the state of equipment. The materials and munitions left over from World War II were not unlimited, nor designed to last forever, and the current level of CMF equipment establishments would prove unsustainable. The ARA also needed re-equipping and, with the level of funds provided by the government, the army focused first on maintaining the equipment of the regulars. The CMF would have to make do with what was left or discarded by the regulars.

From this point on, as materials wore out and were not replaced, the condition of the CMF's share of army equipment worsened. A frequent refrain of citizen soldiers was that they had to spend the first part of their camps repairing the equipment the depot had issued to them. Alex Lochhead of the 58th/32nd Battalion described the equipment with which he practised as sufficient for training but in need of maintenance. C. W. Paterson, who served in 1 RVR and 22 RVR, commented that his units were always short of reliable trucks, and their weapons, especially the M60 machine gun, were often old and incapable of firing. John Barber described the equipment that his 58th/32nd Battalion received as being hand-me-downs, and that trucks and radios were a constant source of problems in the field. Raymond Landgren remembers the shoddiness of the work clothing and the constant breakdowns of the 58th/32nd Battalion mortar platoon's carriers.[14]

Live ammunition also became scarce, and essential munitions were often in short supply. Paterson, an infantrymen, admitted that in his eleven-year CMF career he threw a live grenade only once. At times the situation bordered on the ridiculous. Major Ian Rainford of the 58th/32nd Battalion remembers on more than one occasion bringing fireworks to his unit's camps, to simulate the firing of live ammunition. The citizen soldiers understood the reason: the army had limited resources and, as D. F. Graham commented, the force met the CMF's requirements only after first satisfying the needs of the ARA. As a result, few CMF personnel received training with the army's more advanced equipment.[15]

Perhaps the most worrying development of the 1957 reorganisation was a statement by Menzies instructing the Minister for the Army, John Cramer, to give absolute priority to the build-up, equipping and maintenance of the ARA.[16]

The 3rd Division and Pentropic

While the 1957 modifications to the national service scheme were significant, they did not affect the CMF at its heart, its connection to the community. Yet the shift in focus from the CMF to the ARA foreshadowed future events, policy shifts which proved that the 1957 reorganisation was only the first step in a reassessment of force priorities within the army. The 1957 reorganisation, therefore, marked the beginning of the decline of the CMF and the rise of the ARA.

Despite concern over the shift in policy, it was also true that the government did need to improve the state of the ARA, if the force was to meet its national security obligations. On the eve of the 1957 reorganisation, the ARA's field force units were more than 3700 soldiers under strength, at 54 per cent of establishment out of a requirement of just 8080. Equipment was rapidly ageing and in need of replacement. It soon became clear that the 1957 reorganisation had not gone far enough. The regulars continued to provide considerable assistance to the CMF, both as cadres in citizen force units and as instructors in the national service scheme and had too few soldiers in its field force units to respond rapidly to any Cold War crisis. Therefore, there was a strategic requirement for the regulars to seek, and the government to make, changes in defence arrangements to improve the ARA's abilities. As so often in the past, the necessary improvements would have to come from within the limits of the existing manpower and funding structure. The army had no choice but to pay for the redevelopment of the ARA from within its existing estimate, placing the priority on the ARA at the expense of the CMF. It failed to realise that the nation needed not only an effective regular body but also a viable reserve force.[17]

To fund this internal redistribution, the Military Board embarked on a radical reorganisation of the army's force structure which it hoped would free up the resources the ARA needed. The result was the 1960 Pentropic reorganisation, the most controversial, and ultimately shortsighted, reform the army has ever attempted in its brief history. After just five years, the scheme's failure at all levels was apparent. Not only did Pentropic fail to vitalise the ARA but, in implementing it, the army shattered the fabric of the CMF.

As the name suggests, Pentropic was a divisional structure designed around the principle of five combat elements. Each of these battle groups was based on an oversize battalion, and was to be capable of either fighting alone on an independent mission or joining with other battle groups to form a task force. The Pentropic Division did not contain the traditional nine battalions but instead fielded five large battalions, each approximately 50 per cent larger than its predecessors and consisting of five infantry companies, plus a headquarters and support company. The Pentropic division also contained five field regiments, five field companies, three armoured squadrons and elements of other corps, allowing its commander to provide each task force with a full range of supporting arms.

When the army decided to adopt the Pentropic structure, the ARA's brigade group was still more than 1000 soldiers under strength, and lacked the numbers for the logistic support group it would require upon deployment. The regulars hoped that the Pentropic system would allow it to streamline its rear area, providing more troops for the fighting arms. The army also recommended the suspension of the already truncated national service program, which would relieve the ARA of its instructional duties. By these means, the army hoped to free up over 3800 soldiers. With additional manpower savings from administrative and support echelons, and a reduction in the number of regulars assigned to support the CMF, the force would have enough full-time soldiers to complete the ARA brigade group and raise the needed logistical support.[18]

An additional factor in the conversion to Pentropic was that the United States was also experimenting with the system, and its adoption by Australia would enhance interoperability between the allies. However, the Americans reverted to a traditional triangular division in 1962, leaving the Australian Army as the only force in the world with a five-sided division.

The problems Pentropic posed for the CMF were two-fold. First, when the government ended national service, the CMF lost most of its conscript members. In anticipation of their loss, as well as to reduce the ARA's obligation to the CMF, the army had recently lowered the citizen forces' personnel ceiling to 30,000. Yet resignations greatly exceeded the CMF's ability to retain soldiers and the force's numbers crashed from 51,000 to barely 20,000. Second, to fit its force structure to the new establishment levels, the CMF had to disband or amalgamate many of its units, in all losing 30 infantry battalions, half its artillery, all its infantry brigade and armoured brigade headquarters, and two of its divisional headquarters – the Brisbane-based 1st and the Sydney-based 2nd. Only the 3rd Division remained on the CMF's order of battle. The 1st Division also survived, but as a regular formation.

Even worse for the CMF was Army Headquarters' insistence on redesignating the surviving CMF battalions as State-based regiments. Not only did most communities lose their battalions, but units also lost their regional identities. For example, the army amalgamated the 3rd Division's six-Victorian based battalions into two, 1 RVR and 2 RVR. Outside Victoria, Pentropic battalions bore the names of their respective States.

To the CMF and its supporting communities, battalion names were more than just titles. They embodied the tradition and spirit of generations of citizen soldiers, dating back to their formation during World War I. As with the disbandment of the 39th Battalion during World War II, Army Headquarters jettisoned these familial associations without regard to local sensitivities. By contrast, the ARA was a youthful entity of only thirteen years, and for the regulars tradition had not had time to lay down the roots to support the ethos of its institution. For the CMF the situation was different. Its origins lay in the nation's citizen soldier spirit which first came forth during the colonial period, and then confirmed by the founders of Australia at Federation in 1901. Tradition, continuity and local allegiance mattered to the CMF in ways which the ARA could never appreciate.

Despite a great outcry from those associated with the CMF, the army insisted on the renaming. Brigadier Lowen had no doubt that the army had made a great mistake. 1960 marked the end of the regimental system as the CMF knew it, and he believed that the force lost its way from then on. Ian Rainford, who served in the 58th/32nd Battalion and its Pentropic successor 1 RVR as commander of Essendon Company, believed the Pentropic reorganisation was nothing short of a disaster, and devastated the CMF's infantry, in particular. Rainford pointed out that not only did the citizen soldiers lose 30 infantry units, but the rivalry and esprit de corps that had existed between the old battalions disappeared within the state regiments. Instead, John W. Barber believed, splinter groups appeared within these new units. As a new recruit to 1 RVR in 1960, Barber found the unit divided into a number of cliques based on members' former associations. The overall result for the CMF was a widespread loss of morale and of experienced officers and non-commissioned officers.[19]

In an effort to ameliorate the situation, Army Headquarters did make a small concession to local sensibilities. Instead of designating companies in the Pentropic battalions with anonymous letters, it assigned to them the name of former battalions – a poor compromise that failed to ease the sense of loss felt by citizen soldiers and their associated communities. The issue could become highly emotional. When Ballarat lost its 8th/7th Battalion, the town's mayor retorted that he did not understand 'this 2 RVR nonsense'. He went on to explain that the army failed to realise that 'two generations of men from Ballarat have fought and died in the 8th Battalion'. As a result, the mayor continued, ever since the absorption of the 8th/7th Battalion into 2 RVR, the Ballarat Council had not allowed the successor unit to exercise the Freedom of the City enjoyed by the Ballarat Regiment. The Ballarat council only relented many years later, with the formation of the 8/7 Battalion RVR.[20]

Table 14 outlines the original names of the CMF battalions which made up the five task forces of the Pentropic-structured 3rd Division, and gives their company designations under the new organisation.

Under the Pentropic scheme, the 3rd Division's catchment area expanded to the point where it contained elements from all over the country. In the past, during peacetime, Victoria had been the 3rd Division's traditional heartland, periodically drawing from South Australia and Tasmania. Only in wartime had the army drawn upon the entire nation to constitute the Division. Now, the only regions not in the Division were far north Queensland, Tasmania and the Northern Territory. Table 15 outlines the 3rd Division's order of battle under the Pentropic organisation.

Reinforcing the sense that the 3rd Division's identity now stretched far beyond its traditional Victorian heartland was the wide distribution of its personnel. Table 16 outlines the sources of the division's soldiers under the pentropic system.

The 3rd Division's establishment also included nine civilians, giving the formation a peacetime goal of 13,621. The army had set the wartime establishment of a Pentropic division slightly higher, at 13,962, meaning that the 3rd Division's peacetime establishment was nearly the same as its mobilisation strength. However, throughout the Pentropic period, this was a fiction. The 3rd Division struggled to attract members and remained chronically under-strength.

The wide distribution of the Division under the Pentropic system posed considerable challenges for a citizen soldier-based formation. The key criteria for the CMF had always been that its soldiers and units belonged not just to the army but also to their local communities. From its headquarters in Melbourne, the 3rd Division now had to control units as far away as Perth and Brisbane, a liability for a force whose cohesion was based on local associations. Major-General Kevin Cooke, a future commander of the formation, observed that the Division's 'spread of command was too far geographically and too remote

Table 14
3rd Division Pentropic Battalions
and Renamed Companies[21]

PENTROPIC BATTALION	CMF SOURCE BATTALION	PENTROPIC RIFLE COMPANY NAMES
2 RQR	51st Bn (The Far North Queensland Regt)	A & B Coys – The Far North Queensland Coy
	31st Bn (The Kennedy Regt)	C Coy – The Kennedy Coy
	42nd Bn (The Capricornia Regt)	D & E Coys – The Capricornia Coy
1 RVR	5th Bn (The Victorian Scottish Regt)	A Coy – The Scottish Coy
		B Coy – The Merri Coy
	6th Bn (The Royal Melbourne Regt)	C Coy – The Melbourne Coy
	58th/32nd Bn (The City of Essendon Regt)	D Coy – The Essendon Coy
		E Coy – The Footscray Coy
2 RVR	8th/7th Bn (The North Western Victorian Regt)	A Coy – The Geelong Coy
		B Coy – The Ballarat Coy
	38th Bn (The Northern Victorian Regt)	C Coy – The Sunraysia Coy
	59th Bn (The Hume Regt)	D Coy – The Bendigo Coy
		E Coy – The Goulburn Valley Coy
1 RSAR	27th Bn (The South Australian Scottish Regt)	A Coy – The South East Coy
		B Coy – The River Coy
	43rd/48th Bn (The Hindmarsh Regt)	C Coy – The Mid North Coy
	10th Bn (The Adelaide Rifles)	D Coy – The Adelaide Coy
		E Coy – The Port Adelaide Coy
1 RWAR	11th/44th Bn (The City of Perth Regt)	A Coy – The City of Perth Coy
	16th Bn (The Cameron Highlanders of Western Australia)	B Coy – The Cameron Coy
		C Coy – The Swan Coy
	28th Bn (The Swan Regiment)	D Coy – The West Australian Rifles Coy
		E Coy – The North Coast Coy

personally'.[22] In 1958, Hammer had been able to concentrate most of the 3rd Division at Puckapunyal for his parade. A similar review under the Pentropic structure would have been extremely difficult to organise and practically impossible on financial grounds. As a result of the Pentropic organisation, the 3rd Division's headquarters lost an important function as a nascent field force command, and became more an administrative than an operational body.

As Tables 15 and 16 demonstrate, after the Pentropic reorganisation, the 3rd Division no longer controlled any units in Tasmania. However, 1 RTR was a part of the 3rd Division's

Table 15
Order of Battle – 3rd Division, 1960[23]
Pentropic

TYPE	COMPONENT
Headquarters	3rd Division Headquarters
Armoured	1st/15th Royal New South Wales Lancers
	A Squadron 10th Light Horse
Artillery	Headquarters 3rd Division RAA
	2nd, 3rd, 11th, 13th, 15th Field Regiments
	132nd Division Locating Battery
Engineers	3rd, 11th, 12th, 13th, 14th Field Squadrons
	28th Field Park Squadron
Signals	3rd Division Signals Regiment
Infantry	2 RQR
	1 RVR
	2 RVR
	1 RSAR
	1 RWAR
Intelligence	3rd Division Intelligence Unit
Supply & Transport	Headquarters 3rd Division CRAASC
	10th Transport Company
Medical	4th, 6th, 9th Field Ambulances
Ordnance	3rd Division Ordnance Company
	105th, 107th, 113th Infantry Workshop Stores Section
Electrical & Mechanical	Headquarters 3rd Division CRAEME
	1st/15th Royal New South Wales Lancers LAD
	Squadron Section 10th Light Horse LAD
	2nd, 3rd, 11th, 15th Field Regiment LAD
	3rd Division Signal Regiment LAD
	10th Company RAASC LAD
	105th, 107th, 107th Infantry Workshops

Combat Support Group, a collection of combat and combat support units whose purpose was to augment the Division's strength at mobilisation. While Tasmanian units were not included in the immediate Pentropic organisation, the 3rd Division thus did maintain an association the island State.[24]

Ceremonial Activities

Throughout the Cold War, units of the 3rd Division strove to maintain close ties with the communities upon which they were based, as they had during the inter-war period, highlighting the reality that citizen soldiers were the bridge between two levels of the Australian defence organisation, the local community and the national army.

Participating in local events also had a more practical purpose. By keeping the image of the CMF in front of the community, ceremonial activities also provided a potential

Table 16
Regional Distribution of 3rd Division Personnel
under Pentropic Organisation[25]

COMMAND	ARA ESTABLISHMENT	CMF ESTABLISHMENT	TOTAL ESTABLISHMENT
Northern Command (Queensland)	48	2239	2287
Eastern Command (New South Wales)	16	705	721
Southern Command (Victoria)	138	5406	5544
Central Command (South Australia)	45	2014	2059
Western Command (Western Australia)	55	2946	3001
Total	302	13,310	13,612

forum at which to attract recruits. This was an especially necessary task when conscription was not in effect. Events, however, had to pass muster, and the 3rd Division did not continue to participate in activities which did not provide either a training or recruiting benefit. For example, in 1973, the 3rd Division willingly participated in the Rich River Festival at Echuca. The exercise involved the construction of a footbridge across the river, and a crossing by infantry. However, the Division politely declined to participate the following year, because the exercise failed to produce a single recruit for the Echuca-based C Company, 2 RVR.[26]

The ceremonial event in which the 3rd Division mostly regularly participated was Anzac Day. In 1967 the units of the 3rd Division participated in more than 100 Anzac activities. These ranged from the main Anzac Day march in Melbourne, which attracted hundreds of soldiers, to the contribution of the five other ranks of the 2nd Commando Company who served as guards at the service in Caulfield. In 1967, approximately 1400 soldiers from 3rd Division units took part in one way or another.[27]

Royal visits have always brought the citizen soldiers into the public view. When Queen Elizabeth II toured Australia after her Coronation in 1953, the soldiers of the 3rd Division, with other CMF elements, lined the streets of Melbourne along the route of her procession. Jim Killender, then serving in the 40th Heavy Anti-Aircraft Regiment, spent two full weekends practising drill movements for the event. Taking up his position on the day of Her Majesty's arrival, Killender remembers that the large crowd was in festive mood and generously supplied his unit with tea and biscuits.[28] On another occasion, Graham Farley of the 15th Field Battery participated in the firing of royal salutes to announce a royal birth, and in the visit to Melbourne of the Royal Yacht *Britannia*.[29]

The 3rd Division Post-Pentropic

As early as 1963, the Army's senior ranks had became convinced that the Pentropic organisation was a failure. Its principal fault was that the large size of the battle groups made command and control too difficult for a battalion headquarters. As the division did not have any intermediary headquarters, the function formerly provided by a brigade headquarters, battalions had had to take on responsibility for many more administrative matters than in the past. Moreover, at the strategic level, the great size of the Pentropic battalions limited the number of infantry units the army could constitute within the existing manpower limits, and the ARA simply did not have enough units to meet all its contingencies.

Moreover, Pentropic's lack of compatibility with its allies left the Australian Army in the strange position of being the only force using the model. Had Australia, as a minor power, been called upon to join a coalition, its senior allies would almost certainly have compelled the adoption of a more useful structure. In fact, this is what happened. Every time the army rotated an ARA battalion to Malaya, as part of Australia's commitment to the defence of that region, the chosen unit had to reconfigure itself from the Pentropic structure to something more suited to operations with British forces. On returning home, the unit then had to rebuild itself into a Pentropic form.

On 20 December 1964, the Minister for the Army, Jim Forbes, acting on the recommendation of the Chief of the General Staff, Lieutenant-General Sir John Wilton, announced the demise of the Pentropic system. It now became Wilton's task to untangle the army from the structure and restore it to a more traditional divisional model. As the Pentropic battalion establishments were nearly twice the size of the previous standard, the army used its breakup as an opportunity to raise additional infantry units, reconfiguring the eight State-based CMF battalions into fifteen. In Victoria, 5 and 6 RVR joined 1 and 2 RVR, giving the State four battalions. However there was no restoration of the citizen soldier battalion's former identities. The one concession Wilton made was to agree that the CMF could use the numbers of its former battalions when raising new ones, rather than simply select the next number in the State regiment sequence, as had been originally planned. For example, for the 3rd Division, the army chose the titles 5 RVR and 6 RVR, to promote the association with the former 5th Battalion (The Victorian Scottish) and 6th Battalion (The Royal Melbourne Regiment). The 1965 reform also resulted in the re-raising of the Sydney-based 2nd Division, giving the CMF two divisional headquarters.[30]

The abandonment of the Pentropic system was actually a fairly complex process and led to a considerable contraction of the 3rd Division's regional command, control, and recruitment responsibilities. During the 1965 reorganisation, the army also chose to raise a number of regional ARA and CMF task force headquarters as intermediary headquarters between the Division and the units. Before the Pentropic period, the 3rd Division had contained three brigades (4th, 6th, and 9th) which had provided that function. They contained nine battalions distributed across Victoria, South Australia and Tasmania. After the reorganisation, the 3rd Division had only a single task force, the 4th in Victoria. Wilton authorised the raising of a Headquarters 9th Task Force in South Australia but its commander reported directly to Headquarters Central Command. In Tasmania, the small number of citizen force elements did not justify the creation of a task force headquarters in Hobart, and instead Tasmanian CMF units reported directly to Headquarters Tasmanian Command. Units based in Queensland and Western Australia also became the responsibility of their own local task forces.

The most obvious consequence of the 1965 changes for the 3rd Division was the reduction of its catchment area to just one State, Victoria. Of greater significance, however, was the raising of the task force headquarters, which changed the relationship between the units belonging to 4th Task Force and those of the 3rd Division Headquarters, with implications for command issues upon mobilisation. 4th Task Force Headquarters was responsible for the command and training of field force units under its control. Administrative tasks remained the purview of the 3rd Division Headquarters. Moreover, should mobilisation of the CMF ever occur, the army planned not to despatch divisions in their entirety but their subordinate task forces, or even units, according to readiness and suitability for the mission. Under the task force arrangement, the ability for the 3rd Division to deploy as a division had declined, and instead the ARA would call upon the CMF in piecemeal fashion, incorporating its members into the regular-led operation either as reinforcements or as part of a rotating force. This, of course, reinforced the status of the ARA, making it impossible for the CMF to take the lead in mobilisation and deployment. In effect, the 3rd Division, as a divisional formation, had lost some of its capabilities and importance as a field force command entity.[31]

While the 3rd Division now contained only four battalions, through its 4th Task Force – a relatively small number for a divisional formation – it did gain control of a large number of support and communication zone units, continuing a trend begun in the Pentropic era. The 3rd Division's emphasis moved more towards training and administration rather than command in the field. Training was concentrated on individual or unit skills, and there was no opportunity for the Division to take the field as a formation. Under the Pentropic organisation, the 3rd Division Headquarters had been responsible for a geographic area too vast for effective command. Similarly, after the 1965 reorganisation the Division controlled far too many units, of too many types, to establish an effective presence. The comments of one young soldier are illustrative of this development. Peter Ford joined 2 RVR in 1969, and served in the citizen forces for eleven years. While his junior status – he only reached the rank of corporal – might have distorted his perspective of the army's higher levels, he does not remember ever having been made aware of the 3rd Division's existence during his entire service. The 4th Task Force was the field formation with which his loyalties lay.[32]

Table 17 outlines the organisation of the 3rd Division after the 1964 reform.

Table 17[33]
Order of Battle, Major Units – 3rd Division, 1964[34]

TYPE	COMPONENT
Headquarters	3rd Division Headquarters
Armoured	4th/19th The Prince of Wales's Light Horse
	8th/13th Victorian Mounted Rifles
Artillery	Headquarters Royal Australian Artillery
	2nd Field Regiment
	10th Field Regiment
	10th Medium Regiment
	132nd Divisional Locating Battery
Engineers	3rd Field Engineering Regiment
Signals	3rd Division Signal Regiment
Infantry	Headquarters 4th Task Force
	1 RVR
	2 RVR
	5 RVR
	6 RVR
Medical	4th Field Ambulance
	6th Field Ambulance

As well as those assets, the 3rd Division controlled two attached subordinate headquarters, with numerous sub-units of their own. Headquarters 1st Support Group commanded units such as 3rd Division Column, 6th Transport Column, 11th Base Ordnance Depot, 8th/13th Victorian Mounted Rifles Workshop, 10th Medium Regiment Workshop, 105th Infantry Workshop, and 52nd Communication Zone Workshop. The 6th Construction Group managed a range of corps level engineering support units.[35]

A further blow to the CMF cause was the announcement by Prime Minister Menzies in November 1964 of the restoration of National Service, this time on a selective service basis as a response to the worsening international situation to Australia's north, particularly between Malaya and Indonesia during Confrontation. However, unlike all previous Australian conscription schemes, the government's express purpose was to assist the ARA, not the CMF. This act was a turning point in the allocation of personnel between the two branches of the army. Those selected incurred a two-year, full-time service obligation with the ARA. However, there was some provision for service in the CMF. Those eligible for conscription had the option of instead volunteering for a six-year period of part-time duty with the CMF if they desired to avoid the ARA's full-time service obligation.[36]

Conscription did achieve its objective for the ARA, and the regular army finally had the troops to fulfil its national defence role. However, that success came only at the expense of the CMF. National service did not give the citizen forces access to the full range of

Australia's young men, and made it unable to attract the best and most interested soldiers. The CMF's ranks did swell during this period, but unfortunately only out of the desire of many new recruits to avoid service with the regulars and possible overseas duty in the combat zones of Vietnam, Malaya, and Borneo. Moreover, the government had drafted the legislation poorly and left a loop-hole which allowed an individual to choose the CMF, but then, if his birth date was not selected for conscription, not attend parades and camps, effectively terminating his military liability.

The situation in the 3rd Division is typical of the problems the CMF faced. When the commitment to Vietnam began, the 3rd Division lost many of its more experienced and motivated soldiers, who volunteered for full-time duty and overseas service with the ARA. In exchange, the 3rd Division received disgruntled recruits with little interest in military service. Not surprisingly, the citizen forces gained a reputation as being a haven for draft dodgers, and its status within the community, already alienated by the Pentropic reorganisation, declined further.[37]

The army was not unaware of the problems of the CMF, and in 1966 did approve a limited improvement of the conditions for citizen soldiers. The reduction of 1960 had severely curtailed the CMF's presence in the country, and recruitment in rural areas had fallen. The army decided to raise a CMF battalion in each State with special enlistment terms to appeal to rural recruits. There was, however, a secondary reason. Under the terms of the selective service legislation, men of conscription age had the right to opt for the CMF. However, in rural areas, this was not a viable option as there were no CMF units to join. Thus the raising of special condition battalions was necessary, to provide all men subject to conscription with the same choices. The 3rd Division raised 22 RVR in August 1966. Instead of becoming part of 4th Task Force, the 3rd Division Headquarters commanded the unit directly.

'Special condition' battalions, or 'Bushmens Rifles' as they became known, had a service obligation of 33 days' training a year, which a recruit could fulfil either by attending one camp of continuous instruction for 33 days, or two of sixteen and seventeen days apiece. There was no obligation for nightly parades or weekend bivouacs. The benefit of service with the 'Bushmens Rifles' was that a recruit did not have to live near a CMF centre, as the unit concentrated training into one, or at most two, periods.[38]

Overseas Opportunities: Vietnam

In June 1965, the Australian government dispatched 1 RAR to South Vietnam, as the nation's contribution to the United States' struggle against North Vietnam. This was the first instalment of a contingent which would grow into a three-battalion task force with an overall strength of more than 8000 soldiers. All of the units in Vietnam were part of the ARA (as were those participating in Australia's two other overseas commitments – the Confrontation with Indonesia in Borneo and the ongoing commitment to the defence of the Malayan Peninsula).

Senior CMF officers were well aware that they were missing an opportunity to gain essential operational experience. After the government raised the CMF in 1948, the force could draw upon the great knowledge its World War II veterans possessed. By 1965, most of these men had left the army, and developments in the art of war had rendered the experience of those remaining increasingly obsolete. The CMF Member of the Military Board, Major-General P. A. Cullen, attempted to establish a formal role for the CMF in Vietnam. He suggested the formation of a composite CMF battalion drawn from throughout the force who were willing to volunteer for overseas service. However, the rest of the Military Board rebuffed his suggestion and there was not much support from the government, despite the fact that changes to the *Defence Act* now permitted the despatch overseas of CMF units. The Minister for the Army, Malcolm Fraser, made it clear, however, that he believed there was no place for the CMF in operations such as Vietnam. This he believed was a task for the ARA.[39]

The government's policy left the CMF with two avenues by which members could gain overseas combat experience. The first was either to enlist in the ARA or volunteer for full-time duty with the regulars. According to the commander of 1 RVR at the time, Lieutenant-Colonel Kevin Cooke, many of his best young officers and non-commissioned officers resigned from the CMF and joined the ARA. They included soldiers such as Captain Charles Mollison, who became a company commander in 1 RAR and 6 RAR, and Lieutenant Bill Molloy, who served in 7 RAR. In some cases, a stepdown in rank did not deter those determined to serve. Bill Fogarty was a lieutenant in 1 RVR but went to Vietnam as a corporal in the support company of 7 RAR.[40]

The second method by which CMF officers could serve in Vietnam was via a training program which arranged short-term attachments to ARA units, usually a hectic two to three weeks in theatre with a unit from the same corps as the CMF officer. The total number of 3rd Division officers who spent time on attachment in Vietnam is not known, but, by mid-1968, 31 had availed themselves of the opportunity and another 52 were in the pipeline.[41] The objective of these attachments was more than to provide a few CMF officers with operational experience, or to build a core of leaders with war experience. It was also designed to test CMF training and procedures and to determine whether instruction provided an adequate basis of knowledge for citizen soldier units to take on an operation successfully in the case of mobilisation. These attachments were in no way sops to the CMF; the incoming officers received real experience and participated in all activities of the ARA unit to which they were attached.

A few examples suffice to illustrate. Captain Denis Luttrell came from the Monash University Regiment and spent his nineteen days in Vietnam as an infantry officer in 6 RAR(ANZAC), an Australian battalion which included two New Zealand companies. During his time with 6 RAR, there were two major contacts with the enemy and his company came under heavy fire. In one engagement three Australian soldiers in a nearby position lost their lives. The reality of the situation came home to Everett Reynolds when he went out on his first patrol with 6 RAR(ANZAC). Before departing, he removed his

29: Lieutenant Colonel Kevin Cooke (glasses) CO 1 RVR outside of HQ 7 RAR at Nui Dat, South Vietnam. Surrounding Cooke are former 1 RVR soldiers serving with 7 RAR. They are (back l to r) Sergeant Rowley, Sergeant Gus Barry, Lieutenant Robbie Cathcart, Lieutenant Owen Inman, Lieutenant David Tresize, Corporal Max Hemmings, Corporal Howard, (front l to r) Private Deevers and Lance Corporal Bill Fogarty.
[*Courtesy* Kevin Cooke]

badges of rank, and picked up his share of small arms ammunition, grenades and mines, and prior to leaving the security of the perimeter, with every other man in the patrol, he test-fired his weapon. On another occasion, the patrol he was with had to move across 3000m of ground, including a swamp where the water reached up to his armpits. Major Mike Vincent of 2nd Field Regiment spent his time in Vietnam with the 4th Field Regiment and 2 RAR. One evening, an enemy force attacked the adjoining company's perimeter. An American medium field battery responded in support, and Vincent remembers feeling the concussion from the exploding 155 mm rounds through the side of his weapon pit. Men like Reynolds returned home knowing that a brief tour had made them better CMF officers. They now knew what it was like to have real bullets in their weapons and to face a hostile enemy. By going on patrol, and coming under fire, they gained experience the 3rd Division could not supply in its training camps, and which they could use to improve instruction in future exercises. Their recollections demonstrate that the army expected much of its CMF visitors.[42]

Lieutenant John Stamp, of 3rd Division Engineers, also took advantage of the visiting program, serving in South Vietnam from 15–28 May 1970. He was the sixth engineer from 3rd Division to participate. During his stay he was attached to 1st Field Squadron, 17th Construction Squadron and 1st Australian Logistical Support Group. On his return, he prepared a report on the program and the type of experiences he had had.

12 May: Move by air from Melbourne to Sydney.

13 May: Depart Sydney for South Vietnam via Darwin and Singapore.

14 May: Arrive Saigon; transfer to Nui Dat; report to 1st Field Squadron.

15 May: Attend briefing on current engineering operations; visit Night Defensive Position Elisabeth; attend briefing of 2 RAR on mines; tour Task Force Base Nui Dat.

16 May: Move to Night Defence Base Bridget for attachment to Plant Troop; witness fire fight during night after contact with Viet Cong.

17 May: Take part in road mine clearance task; join helicopter reconnaissance flight; visit American adviser team at Duc Tan.

18 May: Inspect plant and watch mine clearing operation; attend lecture given to 2 RAR on mines.

19 May: Participate in mine clearing operation; while at Night Defence Base Bridget an ambush occurs nearby in which 4 Viet Cong are killed.

20 May: View air strike followed by a company size airborne assault by units of the Army of the Republic of Vietnam; return to Nui Dat; move to 17th Construction Squadron.

21 May: Attend briefing on construction squadron operations; visit engineering works at Baria Hospital; visit four bridge construction sites on Baria-Dat Do Road; join resupply helicopter flight to Long Son Island.

22 May: Inspect Long Son Island Base, move with New Zealand infantry from base to water supply construction project for inspection.

23 May: Return to Nui Dat by resupply helicopter; visit squadron works sites including road works, drainage and quarry.

24 May: Visit defensive position Horseshoe north of Dat Do; move out with land clearing team on a jungle clearing operation including filling in of one small bunker system; move to Nui Dat.

25 May: Air reconnaissance of land clearing tasks; join Plant Troop commander on inspection of various projects; visit 193rd Works detachment.

26 May: Transfer to Vung Tau; join 1st Australian Logistical Support Group and attend briefing; visit various base units.

27 May: Guided tour of Vung Tau and surrounding area.

28 May: Move to Saigon and transfer to Sydney.

29 May: Return to Melbourne.[43]

This outline demonstrates the great amount of activity the Australian Task Force packed into one CMF officer's brief visit. It also shows that supporting arms officers also received attachments of useful and practical relevance to the training of their men at home, and the contribution to the officers' military careers. The tours were realistic, and CMF officers faced the same dangers as the regulars. The war in Vietnam was a fluid situation, and the threat to engineering operations was real. Stamp's report mentions one observed fire fight, a nearby ambush, and other times when other enemy contacts might easily have taken place.

Most importantly, these visits demonstrated that the CMF did have the basic training and ability to participate in field operations in war. Major Bill Hocking, of the 2nd Commando Company, found his time with 2nd Squadron SAS an opportunity to confirm that his training methods in Australia were along the correct lines. Yet 3rd Division officers were realistic in their assessments. Captain Joe Johnson of Headquarters 1st Support Group realised that CMF logistic units had the ability to make the transition to war level operations in a relatively short time; combat arms, he believed, would take longer. This was not the fault of CMF training, but because combat units did not train often enough to gain the necessary cohesion and balance which combat required. Had Cullen been given the chance to put his idea for a CMF battalion into effect, the citizen soldier would have responded to the challenge. The Vietnam War, from the citizen soldiers' perspective, was a missed opportunity. It is a great shame that the government could not have made the effort to find a role for the CMF in the conflict.[44]

The 3rd Division in the 1970s

If the CMF had already under gone considerable pain as a result of a series of reorganisations that had commenced in 1957, the 1970s would provide even greater challenges. In fact, the 1970s were probably the most tumultuous period of change the Australian Army has ever faced.

During those years, the army's leaders implemented major reforms arising from a number of initiatives: the Hassett Report on the modernisation of Army Headquarters and regional commands; the Tange Reforms of the Defence Group; and the contraction of the regular forces after Australia's withdrawal from Vietnam. The army had as well to struggle to attract recruits after Gough Whitlam's suspension of national service in 1972. While all these significant events affected both the ARA and the CMF, the citizen soldiers also had to endure another reorganisation specific to their branch, the Millar Inquiry into the army's reserve system. In the aftermath of all these changes, especially the last named, the CMF disappeared as an institution, and the army disbanded many of its citizen soldier units. In their place, the government authorised a new body called the Army Reserve.

In 1971, the government signalled the future direction of its military policies. The Military Board issued a minute officially redefining the role of the CMF in terms of its provision of support to the ARA. As Cooke points out, it reversed the former policy which had allotted to the CMF the major role in the nation's defence hierarchy.[45] By the close of the 1970s, the full effect of all this turmoil was abundantly clear. As the 1971 Military Board minute had foretold, the citizen soldier branch of the army was a much reduced body, in size and influence, and the ARA was the force's dominant arm.

In 1970, the 3rd Division had responsibility for all CMF units in the Southern Command, the army's regional administrative area, whose borders were more or less contiguous with those of the State of Victoria. Under its control were a great many units, not normally part of a divisional establishment. These included a vast array of engineering and logistical units which in a major conflict would undertake tasks not associated with divisional operations but rather in support of a corps in the communication zone. In addition, 4th Task Force continued to exercise command over most of the formation's infantry elements. Table 18 outlines the 3rd Division's order of battle in 1970.

The 3rd Division's subordinate headquarters contained the following units:

> Headquarters 4th Task Force: 4th/19th The Prince of Wales's Light Horse (plus LAD); 8th/13th Victorian Mounted Rifles (plus LAD); 1 RVR; 2 RVR; 5 RVR; 6 RVR; 1st Independent Rifle Company; and 2nd Commando Company.
>
> Headquarters 1st Support Group: Headquarters 3rd Division CRAASC; Headquarters 3rd Division CRAEME; Headquarters 6th Transport Column; 6th Transport Company (plus workshop); 7th Transport Company (plus workshop); 15th Transport Company (plus workshop); 41st Transport Platoon (amphibian); 33rd Transport Company (plus workshop); 36th Transport Company (plus workshop); 37th Air Dispatch Platoon; 45th Supply Company; 6th Petroleum Platoon; 3rd Air Support Construction Section; 4th Ordnance Field Park; 11th Base Ordnance Depot; 105th Field Workshop (plus stores section); 1st Terminal Ordnance Platoon; 1st Recovery Platoon, 3rd Royal Australian Army Pay Corps Detachment; 3rd Royal Australian Army Nursing Corps Training Unit; and 3rd Women's Royal Australian Army Corps Company.
>
> Headquarters 6th Construction Group: 38th Field Squadron; 29th Plant Squadron; 16th Construction Regiment; 22nd Construction Regiment; 91st Forestry Squadron; 31st Terminal Squadron; and 2nd Movement Control Training Group.
>
> Southern Command Training Group: Melbourne University Regiment; and Monash University Regiment.[46]

The size of the 3rd Division's order of battle is misleading. It suggests a large, well-rounded formation containing all the elements required to wage modern war. However, the size of the order of battle hid serious weaknesses. Most units were greatly under-strength and suffered from poor morale and low levels of readiness. Modern equipment was in short supply, and the CMF had to make do with ARA cast-offs when its units went to camp. One extreme example was the plight of the 91st Forestry Squadron. In 1969, its commander wrote in his annual training report that the unit expected to receive, at last, its sawmill, after a ten-year wait. He observed that the unit's function was to produce sawn and bush timber for the army's engineering requirements, a task which it had not yet had the chance to practise. Instead, his men spent their time at camp practising basic

Table 18
Order of Battle, 3rd Division – 1970[47]

TYPE	COMPONENT
Headquarters	Headquarters 3rd Division
	Headquarters 4th Task Force
	Headquarters 1st Support Group
	Headquarters 6th Construction Group
	Southern Command Training Group
Artillery	Headquarters 3rd Division RAA
	2nd Field Regiment (plus LAD)
	15th Field Regiment (plus LAD)
	10th Medium Regiment (plus LAD)
	132nd Divisional Locating Battery (plus workshop)
Engineers	Headquarters 3rd Division RAE
	2nd Field Squadron
	8th Field Squadron
	10th Field Squadron
	16th Field Squadron
	15th Corps Field Park Squadron
Signals	3rd Division Signal Regiment
	108th Signal Squadron
Infantry	22 RVR
Intelligence	3rd Division Intelligence Unit
Medical & Dental	4th Field Ambulance
	6th Field Ambulance
	2nd Field Medical & Dental Equipment Depot
	3rd Field Hygiene Company
	3rd Dental Unit
Electrical & Mechanical	3rd Field Engineer Regiment LAD

military skills or minor infantry tactics.[48] Another area where the CMF lagged behind the ARA was access to army aviation. Technically, the 3rd Division was supposed to have an organic helicopter unit on its establishment. However, the army never had enough machines to raise the unit, and the CMF was at the bottom of the priority list for access to aviation for training exercises.[49] The 3rd Division was not alone in facing these problems as they were endemic across the CMF. In 1970, the CMF's official establishment was 60,000 but the force had a posted strength of less than 32,000.

Whitlam's suspension of the national service scheme immediately worsened the CMF's personnel problem. National service recruits left in droves, and the CMF had trouble finding replacements in the post-Vietnam environment. By early 1974, the CMF had less than 20,000 soldiers on its rolls and ranks most affected by the decline were private to sergeant. The force had become increasingly top-heavy with leaders, and resembled the nuclear army of the inter-war period. The morale of the citizen forces generally had worsened. Taking participation in Anzac Day celebrations as a barometer of the health of the CMF, in 1967, the 3rd Division enjoyed wide participation across Victoria. However, as the day approached for 1974, the 3rd Division had to advise Headquarters Field Force Command that it anticipated this year's participation would be at an all-time low. The situation was so bad that the 3rd Division's commander, Major-General J. M. McNeill, requested that the army designate the day as a training parade, using the incentive of pay to increase the numbers.[50]

The CMF's leaders were concerned, and they made a number of suggestions to Army Headquarters for the force's improvement. In 1970, N. A. Vickery, a former commander of the 3rd Division, and then serving as the CMF Member of the Military Board, outlined a plan to Major-General Frank Hassett, the chairman of the Army Review Committee. He called for improvement in the conditions of CMF enlistment, more realistic and adventurous training, better access to modern equipment, and a citizen soldier pension scheme to encourage experienced soldiers to re-enlist. Lieutenant-Colonel P. Varma thought the CMF needed better representation at Army Headquarters and suggested the strengthening of the powers of CMF Members of the Military Board. He sought the creation of a parallel citizen soldier structure to oversee training, personnel, logistics and inspections, in effect conceding that the ARA and the CMF were different institutions and required individual solutions. Another senior officer, Major-General P. A. Cullen, was bolder. Calling for the creation of CMF training centres equipped to the same standard that the regulars enjoyed, and the redirection of conscripts to the citizen forces rather than the ARA.[51]

Instead of action, the CMF had to await the recommendations of the 'Committee of Inquiry into the Citizen Military Forces', under the chairmanship of Professor T. B. Millar. The Minister for Defence, Lance Barnard, had authorised the committee's formation in April 1973. It delivered its report the following year.[52] A number of officers submitted ideas on how to improve the CMF. The commander of the 3rd Division, Major-General John McNeill, believed that the army's priorities had shifted too far in favour of the ARA. He recognised that the nation needed regular soldiers, but it also needed a viable citizen soldier force. He called for the creation of a CMF Command which would strive to maintain local units wherever a community expressed an interest, with the goal of reinforcing the connection between the CMF and the community. The new organisation would report directly to Army Headquarters, and would have the same status as the force's other commands. McNeill also wanted the CMF to have the ability to maintain a few units at a higher level of readiness, so that the army could more easily call upon the CMF for overseas deployments. In that request he was echoing an idea of Cullen's, for the raising of a CMF Ready Reaction Force. However, the Millar Report proved to be another lost opportunity. Far from being the CMF's salvation, it was to provide the means of its destruction.[53]

The Committee proposed the merger of the army's various reserve bodies into a single organisation, including several categories of CMF service and the ARA's existing reserve bodies, the Regular Army Reserve, the Regular Army Emergency Reserve, and the retired list. The new organisation was to be the Army Reserve, a designation which gave little doubt of its purpose: to support the regular army. The adoption of the name 'Army Reserve' served to reinforce the changing relationship between the two halves of the army which had been under way since 1957. When the Australian Army had come into existence in 1901, the primary task of the permanent soldiers (the ARA's distant predecessors) was to assist the citizen soldiers who were the primary defenders of the nation. The ARA was now the dominant partner, and the reserves were there to support the regulars when called upon.

The CMF recognised the significance of the name change, and its officers canvassed other options, such as the 'Australian National Guard' or 'Australian Territorial Army'. T. K. Halliday of the 3rd Division Headquarters staff put forward the suggestion,'Australian National Defence Force'. All these terms suggested a regional focus and, as one commentator opined, 'implies citizen participation in military affairs'.[54] The commander of the 3rd Division, Major-General K. D. Green, wrote to the CMF Member of the Military Board, Major-General A. C. Murchison, to lobby for the citizen soldiers. He pointed out that the term CMF was a household name and readily understood by the public. He continued that a name change would only make recruitment more difficult as the citizen soldiers would have to create a new image. As a last effort he suggested that the army at least postpone the name change until recruitment had improved. Major-General J. M. McNeill, the 3rd Division's next commander, also took up the cause. He wrote:

> I personally see no reason to change the title of Citizen Military Forces, it seems correctly descriptive, it is well known, I don't know anything better suited nor can I see any real necessity for a change.[55]

Their efforts, and those of others had no visible effect on the decision-making process and the name change went ahead.[56]

The Millar Committee recommendation which had the greatest effect on the CMF was the proposal that the army should only retain on its order of battle those citizen force units which were able to maintain near-establishment strengths. Millar suggested that a posted strength of 70 per cent of establishment be the cut-off. This was a dramatic departure from past practice in Australia. Previously, the army had accepted that without conscription it was necessary to tolerate weak CMF units, which nevertheless provided the nation with an expansionary base and a pool of trained leaders. However, from the formation of the Army Reserve, the reserves would no longer be the template for expansion. Instead, the army expected the reserves to fit into the ARA's structure.

Furthermore, Army Headquarters decided to use ARA establishments as the criteria, not the usually lower CMF levels. This meant that Army Headquarters planned to assess already weak units against a tougher criteria than those the force had previously used. In

1970, the 3rd Division's 2 RVR had an establishment of 538 and a posted strength of 345 all ranks. This meant that the unit was at 64.1 per cent of establishment. While below the cut-off, this was close enough that a determined recruitment drive could have saved the unit. However, the army insisted that CMF units had to meet the more exacting ARA establishment, which, in 2 RVR's case, increased its manpower goal to 792, lowering its percentage of establishment figure to just 43.5 per cent. Further compounding the battalion's situation was that after the suspension of national service, the unit's membership fell precipitously. By the time, the army assessed 2 RVR, its strength had sunk to just 183 all ranks, or only 23.1 per cent of target. The 70 per cent cut-off was far too a distant goal for the unit to reach.[57]

The application of this standard to the CMF also ignored the fact that the army had traditionally set citizen soldier establishments on an individual basis, depending on the ability of the unit's catchment area to provide recruits. CMF establishments could vary from unit to unit, even in the same corps, depending on the conditions within their communities. The only periods in the nation's history during which the CMF maintained its units at mobilisation strength were during the world wars or in periods of conscription. At other times, the army had always expected, and permitted, citizen forces to maintain their units below wartime establishments. The sudden change of policy greatly exacerbated the difficulties the units faced in meeting the arbitrary 70 per cent threshold. In the end most failed, and widespread disbandments began to flow through the CMF – now the Army Reserve – by the end of 1974.

The effect on the 3rd Division encapsulates the trauma suffered by the CMF during restructuring. The evaluation of CMF units began in 1974, and the 3rd Division implemented the results the following year. Table 19 outlines the fate of the 3rd Division's major combat units.

Table 19
Reorganisation of 3rd Division, Major Combat Units – 1974[58]

UNIT	ESTABLISHMENT GOAL	ACTUAL STRENGTH IN 1974	FATE
2nd Field Regiment	475	85	amalgamated as 2nd/15th Field Regiment
15th Field Regiment	475	121	amalgamated as 2nd/15th Field Regiment
1 RVR	792	158	amalgamated as 1 RVR
5 RVR	792	161	amalgamated as 1 RVR
6 RVR	792	91	amalgamated as 1 RVR
2 RVR	792	183	amalgamated as 2 RVR
22 RVR	792	90	amalgamated as 2 RVR
1st Independent Rifle Company	259	49	amalgamated as 2 RVR

In a similar manner, the army also amalgamated or disbanded the 3rd Division's support arms. Across the CMF, Hassett oversaw the elimination of 17 major and 224 minor units.

In addition to the loss of many of its units, the 3rd Division also suffered a reduction in status. In February 1976, as one of the Millar recommendations, the army amalgamated Headquarters 4th Task Force and Headquarters 3rd Division to form Headquarters 3rd Division Field Force Group. At the same time the army downgraded the rank of the formation's commander from major-general to brigadier. The army imposed the lower rank because the size of the resulting formation did not warrant a major-general position. With only two battalions on its order of battle, the 3rd Division was little more than a brigade group in size. In addition, Headquarters 3rd Division Field Force Group was to control its subordinate units directly. Under the new arrangement there was no intermediary level of command, such as Headquarters 4th Brigade or 4th Task Force. The 3rd Division Field Force Group's superior headquarters was Field Force Command in Sydney. Table 20 outlines the order of battle of the 3rd Division Field Force Group in 1978.[59]

Table 20
Order of Battle – 3rd Division Field Force Group, 1978[60]

TYPE	COMPONENT
Headquarters	Headquarters 3rd Division Field Force Group
Armour	4th/19th The Prince of Wales's Light Horse
	A Squadron 8th/13th Victorian Mounted Rifles
Artillery	2nd/15th Field Regiment
	10th Medium Regiment
	132nd Divisional Locating Battery
Signals	108th Signal Squadron
Infantry	1 RVR
	2 RVR
	2nd Commando Company
Intelligence	3rd Division Field Force Group Intelligence Unit
Transport	7th Transport Company
	15th Transport Company
Ordnance	3rd Ordnance Platoon
Medical & Dental	6th Field Ambulance
	3rd Preventive Medical Company
	3rd Dental Unit
	3rd Psychology Unit
Pay	3rd Royal Australian Army Pay Corps Detachment

Under its subordinate headquarters for 6th Construction Group, the 3rd Division Field Force Group continued to control a number of corps level engineering units. These included:

7th Field Engineering Regiment
91st Forestry Squadron
39th Electrical & Mechanical Squadron
203rd Works Section
Headquarters 22nd Construction Regiment
105th, 106th Construction Squadrons
107th Plant Squadron.[61]

The task given to the renamed formation was that of commanding all army reserve field force units in the 3rd Military District (Victoria), including overseeing unit training, personnel and administration. Missing, however, was any adequate definition of a role for the formation within national defence policy. The function of the Army Reserve was now to support the ARA, but how this was to occur was not clear, and key issues such as probable missions, operational theatres, and training goals remained unspecified.

Like the Pentropic reorganisation, the implementation of the Millar recommendations caused tremendous pain across the CMF. Certainly some pain was unavoidable. By 1970, the CMF was a force swollen with too many inefficient units too far under-strength to be of any use in an emergency, without a considerable period of building up. After 1972, the CMF's situation had grown worse, with national service optees deserting the institution. Therefore, reform was necessary for the CMF to play any role in the defence of the nation.

However, the real problem with the CMF's reform lay in the manner of its implementation. The ARA showed little sensitivity or understanding of the effect that wholesale elimination of units would have on the CMF. More significantly, the army then did not move forward with positive steps that would revitalise the morale of the citizen soldiers and lay the basis for the creation of a viable Army Reserve, one that was a partner of the ARA, not a neglected subordinate. One of the recommendations Millar had made was for the integration of the regular and reserve branches into a total force. This did not occur, and the ARA saw the reform more as a way to subsume the reserves instead of helping them to grow.

The 3rd Division in the 1980s

While the 3rd Division Field Force Group did not continue to decline in the 1980s, neither did it improve. Once the instability of the Millar reforms came to an end, the CMF entered a period of stagnation. The formation continued to command a diverse range of units, but its size remained that of a brigade group, not a division. Table 21 outlines the 3rd Division's order of battle as it existed in 1980.

Table 21
Order of Battle – 3rd Division Field Force Group, 1980[62]

TYPE	COMPONENT
Headquarters	Headquarters 3rd Division Field Force Group
Armour	4th/19th The Prince of Wales's Light Horse
Artillery	2nd/15th Field Battery
	10th Medium Battery
	132nd Divisional Locating Battery
Signals	108th Signal Squadron
Infantry	1 RVR
	2 RVR
	2nd Commando Company (plus 126th Signal Squadron)
Intelligence	3rd Division Field Force Group Intelligence Unit
Transport	7th Transport Company
	15th Transport Company
Medical & Dental	6th Field Ambulance
	3rd Preventive Medical Company
	3rd Dental Unit
Ordnance	3rd Ordnance Platoon
	3rd Ordnance Service Unit
Electrical & Mechanical	3rd Electrical & Mechanical Service Unit
Pay	3rd Royal Australian Army Pay Corps Detachment
Psychology	3rd Psychology Unit

The Headquarters 3rd Division Field Force Group also continued to control Headquarters 6th Construction Group. The units under its command were: 22nd Construction Regiment, 7th Field Engineer Regiment, 91st Forestry Squadron, 39th Electrical & Mechanical Squadron, and 203rd Works Section.

The main problem in this period for the Army Reserve and the 3rd Division Field Force Group was the army's difficulty in identifying a role for its part-time soldiers. In part, Army Headquarters was not to blame since the force's political masters also largely failed to articulate a role for the army as a whole. After the withdrawal from Vietnam, the government's national security policy focused on the continental defence of the Australian mainland. Since the Whitlam Ministry and up to the 1990s, successive governments have had little interest in overseas operations. Under the policy of continental defence, the Royal Australian Navy and Royal Australian Air Force had the task of controlling the maritime environment to the continent's north. It was the job of these services to make sure an opponent was unable to gain a lodgement on Australia's shores.

Neither the government nor the army could work out how to fit ground troops into the concept of continental defence. The force's leaders in Canberra worked in a policy vacuum, and can perhaps be excused for neglecting the direction of the Army Reserve.

The Army therefore tended to work from the government's previous policy, and sought to maintain a capability to mount overseas expeditions for any regional Cold War conflict to which Australia might make a contribution, and to maintain an expansion base in case of a more major war. As a result, throughout the period, a large gap existed between what the government assumed to be the nature of future conflict and the structure that the army actually maintained.

In the first half of the 1980s, the government and the army did make some efforts to improve the Army Reserve, and achieve the concept of a total force that Millar had conceived. However, the initiatives were limited, and decision makers failed to carry them through. Most ideas required the commitment of additional monies to the Army Reserve and, as so often in the past, the reluctance of the Treasury to provide the military with the resources it required inhibited the force's development. Some government initiatives were actually counter-productive. For example, in 1983 the government resumed taxing reserve pay, eliminating one of the few incentives the Army Reserve had to attract recruits. In another instance, in 1988, the government liberalised the call-out provisions of the *Defence Act*, simplifying the ability to mobilise the reserves. However, the legislation did not include any protection for a reservist's civilian employment, an essential requirement which would have made reserve service much more attractive.

One positive change, however, was the reactivation in April 1984 of the 3rd Division Headquarters. At a ceremony in the Division's headquarters building, 'Grosvenor', the formation's commander, Major-General Kevin Cooke, broke out the old 3rd Division flag to celebrate the occasion.[63]

It was not until 1985 that the government made a real attempt to define a role for the army which would accommodate the force within continental defence. That year, the Minister for Defence, Kim Beazley, asked the consultant Paul Dibb to make recommendations on the composition of the nation's defence forces. Dibb presented his report in 1986 and Beazley incorporated many of its suggestions into the 1987 White Paper *The Defence of Australia*. Dibb pointed out that at present there was no agreement between the government and the Australian Defence Force on what the nation's armed forces were structured to do. He drew attention to the fact that the Army Reserve, in particular, suffered from a lack of purpose and direction, and commented on the foolishness of the government's recent reimposition of tax on reservists' pay.

Yet Dibb's most important recommendations concerned the nature of future war and the role of the Australian Army in its waging. He believed that the army's organisation was too heavy, and remained inappropriately oriented to the provision of an expeditionary force or mobilisation for a major conflict. Instead, he believed, the army should prepare itself for low-intensity warfare in the north of Australia, should enemy forces succeed in crossing the air-sea gap. Dibb expected that Australia's navy and airforce would largely control the nation's maritime approaches. In effect, this meant that the army needed only to prepare for small, widely dispersed landings. Dibb called for the force's reorganisation into a lighter, more mobile and flexible structure, composed of field force units with the

30: 105 mm howitzer in full recoil during the 1981 camp of the 2nd/15th Field Regiment at Puckapunyal. [*Courtesy* Jim Killender]

ability to operate independently in dispersed mode across the country's north. To fulfil this function he also saw a need for the army to move more of its regular combat forces northward to Darwin, where they could train and acclimatise themselves to their anticipated area of operations.[64]

Dibb's findings had serious implications for the Army Reserve. One of the longstanding roles for the nation's citizen soldiers was to provide an expansion base in case of major threat. This was part of the reason why the army maintained under-strength divisions, such as the 3rd, on its rather small order of battle. However, the army would now have difficulty justifying divisional structures if the only threat it had to prepare for was low-intensity warfare and the defence of installations. Asset security was a vital task if Australia was to secure the nation's maritime approaches, but, it was a static garrison role, traditionally given by an army to its second-echelon troops. This task drove home the point that the

31: The restoration of the 3rd Division Flag in April 1984.
Major General Kevin Cooke (in uniform) with former commanders of the 3rd Division (*l to r*) Major General R. R. Gordon, Major General N. A. Vickery and Major General K. D. Green. [*Courtesy* Kevin Cooke]

regular troops were the ones who would enter future battle as a field force, while the Army Reserve assumed the rear area security detail.

To fulfil these new roles, the army gave the 3rd Division, and the South Australian based 9th Brigade, the task of asset protection. The 3rd Division's two final commanders, Major-Generals B. N. Nunn and W. E. Glenny, organised a number of exercises in Darwin, Curtin, Tindal, Kununurra and Wyndham. In 1989, the 3rd Division participated in the army's major exercise, Kangaroo 89, and more than 3000 members of the 3rd Division exercised in the far north.[65]

A consequence of Dibb's recommendation to move the ARA northwards was that it further isolated the reserves from their regular counterparts. Moreover, it also decreased the army's visible presence in the communities of the more populous south. The move to

Darwin of 1st Brigade, which commenced in 1991, made operational sense, but removed much of the army from the nation's heartland. As the army shifted northwards, the Army Reserve's centre of gravity similarly moved north. Consequently, the status of Victorian-based reserve units, now among the force's most distant, declined, while those in Queensland grew in importance.

Before the 3rd Division could master its new role, events at the national defence policy level determined the formation's fate. In 1991, the Department of Defence presented an efficiency report, the 'Force Structure Review'.[66] The Minister for Defence, Robert Ray, faced a difficult problem. He had to decide how to pay for a significant expansion of Australian naval and airforce power, including the purchase of new and expensive platforms for those services. The army was also to receive some new equipment, such as helicopters and vehicles, but its shopping list was relatively affordable. The problem facing the services was Treasury policy, which insisted that the services pay for any additional defence expenditure from their existing votes. The army, in particular, would bear the largest part of the burden but would have to make do with little or no growth in its estimate for the foreseeable future. The implication of this fiscal policy was that, to free up money for necessary capital expenditures, the army had to implement a major contraction of its force structure and reduce its establishment by more than 5000 uniformed personnel.

The confluence of the recent changes in strategic policy and the need for fiscal efficiency brought about the disbandment of the 3rd Division. The Chief of the General Staff, Lieutenant-General John Coates, decided that the army no longer had the luxury of being able to maintain divisional formations. In June 1991, the army removed the 3rd Division from its order of battle. Its successor organisation was the Victorian-based 4th Brigade. Headquarters 4th Brigade took over the 3rd Division's remaining units, and responsibility for asset protection in the far north. Table 22 outlines the units under its control.

Table 22
Order of Battle, 4th Brigade, 1991[67]

TYPE	COMPONENT
Headquarters	Headquarters 4th Brigade
Artillery	2nd/10th Medium Regiment
Engineers	10th Field Squadron
Signals	108th Signal Squadron
Infantry	5/6 RVR
	8/7 RVR
Transport	7th Transport Company
Ordnance	4th Field Supply Company
Electrical & Mechanical	105th Field Workshop
Medical	6th Field Ambulance
	3rd Field Dental Unit

The fate of the army's other divisions was slightly different but essentially their ability to take the field as divisional formations also came to an end. Coates reorganised the Brisbane-based and ARA-manned 1st Division into a deployable Joint Force Headquarters. In Sydney, the 2nd Division remained on the army's order of battle but only as a study cell, and had no field force capability. After 75 years of service to the nation, the 3rd Division had ceased to exist.

Endotes

1 'Army Post War Plan', March 1947, AA(ACT), A816/1, item 521/301/245.
2 K. G. Cooke, 'One Army', in Peter Dennis and Jeffrey Grey, eds., *The Second Fifty Years: The Australian Army, 1947-1997* (Canberra: Australian Defence Force Academy, 1997), pp. 71-90.
3 *See* 'Designation of Units of the Post War Army', 26 February 1948, AA(ACT), A2653/1, 1947, volume 4.
4 Information taken from 'Post War Army Organization', (1948), AWM113, item MH1/175; and 'Designation of Units of the Post War Army', 26 February 1948, AA(ACT), A2653/1, 1947, volume 4.
5 D. S. McCarthy, 'The Once and Future Army: An Organisational, Political and Social History of the Citizen Military Forces, 1947–1974' (Thesis submitted for the degree of Doctor of Philosophy in the School of History, University College, Australian Defence Force Academy, University of New South Wales, 1997), pp. 30-33; and information provided by Colin H. Brown.
6 Information provided by Robert N. Thompson, Alex Lochhead and R. F. C. Campbell.
7 Information provided by I. H. Lowen.
8 'Unpublished Memoir of R. R. Gordon' (author's Possession).
9 J. E. Murphy, 'History of the Post-War Army', ADFA Library, 146233, pp. 57-79.
10 Information taken from parade program, 'Third Division Parade at Puckapunyal, 23 March 1958' (author's possession).
11 Information taken from exercise program, 'Operation Firepower' (author's possession).
12 Jim Killender, 'Wanted but Never Needed: Stories of My Experiences as an Officer in the Weekend Warriors' (unpublished manuscript, author's possession); and information provided by I. H. Lowen.
13 Information taken from exercise program, 'Operation Firepower' (author's possession).
14 Information provided by Alex Lochhead, C. W. Paterson, John Barber and Raymond Landgren.
15 Information provided by C. W. Paterson, Ian Rainford and D. F. Graham.
16 'Minutes of Conference Held at Victoria Barracks', 31 May 1957, AA(ACT), A6059/2, item 41/441/47; and McCarthy, 'The Once and Future Army', p. 69.
17 'Minutes of the Reorganisation Conference', 12 March 1959, AA(ACT), A6922/1, item 1/14.
18 Figures taken from 'Current and Planned Allocation of Strengths, Australian Regular Army and Civilians', 21 December 1959, AA(ACT), A6059/2, item 41/441/69, part 1.
19 Information provided by I. H. Lowen, Ian Rainford, John W. Barber and A. J. Peter Nattrass.
20 Quoted in Cooke, 'One Army', pp. 76-77.
21 Information taken from AA(ACT), A6059/2, item 41/441/69, part 2.
22 Cooke, 'One Army', p. 77.
23 Information taken from 'The Reorganisation of the AMF 1959/1960-1961/1962: Planning Order of Battle', 15 November 1960, AA(ACT), A6922/1, item 1/14.
24 For the organisation of the 3rd Division Combat Support Group *see*: 'The Reorganisation of the AMF 1959/1960-1961/1962: Planning Order of Battle', 15 November 1960, AA(ACT), A6922/1, item 1/14.
25 Information taken from 'Army Planning Orbat – Summary', 10 March 1962, AA(ACT), A6922/1, item 1/14.
26 'Echuca Rich River Festival', 29 March 1974, AA(VIC), B420/1, item R274/74/6.

27 'Anzac Day Commemoration Activities 1967', Victoria Barracks, Melbourne, MP1481, item R274/1/16, part 1.
28 Killender, 'Wanted but Never Needed', p. 20.
29 Information provided by Graham Farley.
30 'New Organisation of the Australian Army – Statement by Minister for the Army, Dr. the Hon. A. J. Forbes', 20 December 1964, and 'Re-organisation of the CMF', AA(ACT), A6922/1, item 1/29.
31 'MBI No. 1-4: Intermediate HQ Function', November 1965, AA(ACT), A3688/25, item 605/R2122/2.
32 Information provided by Peter Ford.
33 Information taken from 'Allocation of Major Units to Formations', AA(ACT), A6922/1, item 1/29.
34 Information taken from 'Post War Army Organization', (1948), AWM113, item MH1/175; and 'Designation of Units of the Post War Army', 26 February 1948, AA(ACT), A2653/1, 1947, vol. 4.
35 *Ibid.*
36 Peter Edwards, *Crises and Commitments: The Politics and Diplomacy of Australia's Involvement in South East Asian Conflicts, 1948-1965* (North Sydney: Allen & Unwin, 1992), pp. 271-72, 329.
37 Grey, *A Military History of Australia*, p. 233; and Cooke, 'One Army', p. 77.
38 For the story of the special condition battalions *see*, Neil Leckie, *Bushmens Rifles: A History of the 22nd Battalion, The Royal Victoria Regiment, 1966-1975* (Mulgrave, VIC: The Ex-22nd Bn RVR Association, 1999).
39 McCarthy, 'Once and Future Army', pp. 224-27.
40 Information provided by Kevin Cooke.
41 'Visits by CMF Officers to Vietnam', 17 June 1968, Victoria Barracks, Melbourne, MP1481, item R875/1/4.
42 Information provided by Denis Luttrell, Everett Reynolds and Mike Vincent.
43 Information taken from 'Visits by CMF Officers to South Vietnam: Officers Report', 2 November 1970, in 'History of the Royal Australian Engineers, 3rd Division CMF' (unpublished manuscript, author's possession).
44 Information provided by Bill Hocking and Joe Johnson.
45 Cooke, 'One Army', p. 78.
46 Information taken from 'CMF Units in Being', 25 March 1970, AA(ACT), A3688/26, item 174/R1/98.
47 *Ibid.*
48 'Annual Training Report – 91st Forestry Squadron', 21 May 1969, Victoria Barracks, Melbourne, MP 1481, item R810-8-5.
49 'Director Army Aviation to Director Staff Duties, re Army Aviation for CMF Camps', 9 April 1970, AA(ACT), A3688/34, item 174/R1/82.
50 '3rd Division Routine Message', 7 April 1974, AA(VIC), B420/1, item R274/74/1.
51 'Army Review Committee – Submission by Major-General P. A. Cullen re: Citizen Military Forces', 5 May 1970, and N. A. Vickery, 'The Decline in Strength of the CMF', 30 September 1970, AA(ACT), A3688, item 582/R1/20; and P. Varma, 'The CMF Member's Secretariat', 7 December 1972, AA(VIC), B420/1, item R174/1/7.
52 For the committee's report see *Report of the Committee of Inquiry into the Citizen Military Forces* (Canberra: AGPS, 1974).
53 'Submission by Major General J. M. McNeill', 2 July 1973, AA(VIC), B420/1, item R174/1/7; and 'CGS to Minister for Defence, CGS no. 185/1973', June 1973, AA(ACT), A6835/1, item 4.
54 'Renaming of CMF', AA(VIC), B420/1, item R174/1/6.
55 'Submission by J. M. McNeill to the Committee of Inquiry into the CMF', AA(VIC), B420/1, item R174/1/7.
56 'Renaming of CMF', and 'Green to Murchison', 28 February 1973, AA(VIC), B420/1, item R174/1/6; and 'T. K. Halliday to HQ 3rd Division', February 1973, AA(VIC), B420/1, item R174/1/7.
57 'CMF Units in Being', 25 March 1970, AA(ACT), A3688/26, item 174/R1/98; and 'CGS to the Minister for Defence, CGS no. 366/1974', AA(ACT), A6835/1, item 5.
58 Data taken from 'CGS to the Minister for Defence, CGS no. 366/1974', AA(ACT), A6835/1, item 5.
59 *Report of the Committee of Inquiry into the Citizen Military Forces*, pp. 69-70.

60 Information taken from 'Effective Unit Strengths as at 30 June 1978', AA(ACT), B420/1, item R211/1/23.

61 *Ibid.*

62 Information taken from 'Outline of 3rd Division' (author's possession).

63 Information provided by Kevin Cooke.

64 Paul Dibb, *Review of Australia's Defence Capabilities* (Canberra: AGPS, 1986), pp. vi, 15, 26-27; see also Cooke, 'One Army', pp. 80-81; and Department of Defence, *The Defence of Australia* (Canberra: AGPS, 1987).

65 *Pipeclay, The Magazine of the Melbourne Naval & Military Club* (1991), p. 3.

66 *See Force Structure Review* (Canberra: AGPS, 1991).

67 'Land Command Force Structure Review', 17 June 1991, Defence Central Registry, A6721, item A90-2829, part 8.

Chapter 6
The Legacy of the 3rd Australian Division

On 29 June 1991, at the Melbourne Town Hall, Major-General Kevin Cooke rose to address a gathering of the members of the 3rd Division.[1] Before him were nearly 300 former and currently serving citizen soldiers who had had an association with the Division. Several years earlier, planning for the occasion had begun in anticipation of the happy event of the 3rd Division's 75th Anniversary. Sadly, as Cooke looked out at his audience, everyone was aware that the evening had recently gained another purpose. It was the final meeting of 3rd Division comrades, brought together one last time on the eve of the formation's disbandment.

The 3rd Division was the country's longest serving division, and its story corresponded with much of the history of the Australian Army and the Australian nation. It served with distinction in World Wars I and II, and helped our allies secure victory over determined and dangerous foes. The 3rd Division's battle honours were considerable and well deserved. They range from the formation's lonely stand against the German onslaught before Amiens, in the darkest days of 1918 when all appeared lost, to the tenacious battle against the Japanese in the rugged isolation of New Guinea during the campaign for Salamaua. In all its encounters with the enemy, the 3rd Division excelled and proved that citizen soldiers have the ability to come to Australia's defence in the moment of need.

If the 3rd Division's war history is filled with glorious and heroic deeds, the longer years of peacetime provided different and perhaps more difficult challenges. Between the two world wars, citizen soldiers had to contend with an indifferent government which abdicated the responsibility for national security and wanted to avoid, as far as possible, paying any of the costs of the Commonwealth's defence. The result was an army so starved of funds that it virtually disappeared. That the 3rd Division survived the era at all was only a result of the determination of its citizen soldiers to keep the institution alive and was a testament of their loyalty to Australia.

Given the government's inter-war security policy, when war came again in 1939, the 3rd Division was impotent, unready for combat at any level. Yet, while it could not take the field, it had fulfilled its other task, to maintain a cadre of leaders and men who could lay the foundation for the 2nd AIF. When Japan threatened Australia directly in 1942, the 3rd Division again came forward, becoming one of the nation's field formations, and securing new battle honours in the campaigns of New Guinea and Bougainville.

32: Major General Kevin Cooke toasts the 3rd Division at Melbourne Town Hall on the eve of its disbandment. [*Courtesy* HQ 3rd Division]

In the aftermath of World War II, the 3rd Division faced new challenges. Once again, its leaders had to contend with successive governments which vacillated between moderate interest and total indifference to defence issues. More important for the citizen soldiers were the post-war changes in Australia's security environment. After the Singapore debacle, the Commonwealth's political leaders accepted that Australia could no longer rely on the Empire for the nation's defence. Australia had to accept greater responsibility for its own security. The need to provide for defence, and to discharge Australia's substantial post-war military responsibilities, directly led to the creation of the nation's first standing field force, the ARA. The formation of the ARA was a turning point in the military history of Australia, since in the past, the citizen soldier had been the sole representation of the nation's martial spirit.

33: 75th Anniversary and Disbandment Dinner, 29 June 1991, Melbourne Town Hall.
[*Courtesy* HQ 3rd Division]

In addition to perceiving a need for regular soldiers, Australia's political leaders also knew that they now had to cultivate a new ally, for the nation's protection against a major threat. That protector became the United States. However, external guarantees of security do not come without a price. In the first half of the century, Australia had come to the aid of the Empire in two world wars. American security guarantees led to Australian participation in the Korean, Vietnam and Gulf Wars.

As the post-World War II era began, the relationship between the branches of the Australian Army was no different to what had existed in the past. The permanent soldiers had gained a field force capability, but the ARA remained below the CMF in the nation's defence hierarchy, and the government recognised citizen soldiers as the army's primary branch. From 1957, however, the relative power balance within the army, and the government's perception of defence requirements, began to change. Gradually, the CMF's status declined, reaching a point where the citizen soldiers even lost their cherished name to the more subservient term Army Reserve. At the same time the ARA rose in power and became the force's dominant element.

Although spread over a period of more than thirty years, the changes were deeply traumatic and emotional for most citizen soldiers. Some went so far as to suggest that the

CMF's decline was a direct result of deliberate ARA policy. Such suggestions go too far, and fail to recognise that the changes in the Commonwealth's security environment during the Cold War mandated a greater reliance on regular forces. However, the ARA and the government do have a case to answer.

If change was necessary, as it was, it could have been achieved in a less callous and mean-spirited manner. The broader army gained nothing by the regulars' consistent disregard of CMF history, tradition and culture. The government's failure to clearly enunciate policy, identify roles for its military forces, and resist its traditional desire to limit spending on defence fostered an 'either/or' mentality at Army Headquarters which did both the CMF and the ARA a disservice. Unfortunately, when confronted with choices, the army's leaders invariably chose to provide for the ARA first, and the CMF a distant second.

This need not have been the path taken. The history of the 3rd Division shows that citizen soldiers are important in Australia's defence hierarchy. However, they require a defence administration willing to nurture its citizen soldier organisation and recognise its special needs. The repeated vacillations over tax on the pay of part-time soldiers suggest a government more focused on fiscal policy than the needs of its soldiers. It has taken the Australian government 100 years to provide its citizen soldiers with any job security for their civilian employment on mobilisation; a guarantee was not enshrined in legislation until 2001.

The government and, since 1948, the regular army, have failed to recognise that benign neglect of the nation's citizen soldiers has brought only decay and inefficiency. Defence leaders must provide the Army Reserve, the current form of Australia's citizen soldiers, with the support, resources and sympathy it needs. The past suggests that the time will come again when the security of Australia requires the nation's citizen soldiers to come forward as their predecessors did. Hopefully, then, the Commonwealth will again have a vibrant pool of citizen soldiers, who (while not trained to the same level as the regulars), can draw upon the history, tradition and experiences of formations such as the 3rd Division. Then, they too will become 'Defenders of Australia'.

Endnote

1 'Address by Major General K. G. Cooke at a Dinner Held at the Melbourne Town Hall on Saturday, 29th June, 1991 to Mark the 75th Anniversary and the Disbandment of the 3rd Division' (author's possession).

Appendixes

Appendix 1
Commanders of the 3rd Australian Division

Major-General Sir John Monash	1916–18
Major-General Sir John Gellibrand	1918–22
Major-General G. J. Johnston	1922–27
Major-General H. E. Elliot	1927–31
Major-General Sir Thomas Blamey	1931–37
Major-General E. A. Drake-Brockman	1937–42
Major-General Sir Stanley G. Savige	1942–44
Major-General Sir William Bridgeford	1944–47
Major-General Sir George Wootten	1947–50
Major-General S. H. W. C. Porter	1950–53
Major-General Sir Robert J. H. Risson	1953–56
Major-General H. H. Hammer	1956–59
Major-General N. W. Simpson	1959–60
Major-General R. R. Gordon	1960–63
Major-General N. A. Vickery	1963–66
Major-General S. M. McDonald	1966–70
Major-General K. D. Green	1970–73
Major-General J. M. McNeill	1973–76
Brigadier W. H. Grant*	1976–77
Brigadier J. E. Barry*	1977–80
Brigadier N. E. Bavington*	1980–81
Major-General K. G. Cooke*	1981–84
Major-General K. G. Cooke	1984–85
Major-General J. E. Barry	1985–87
Major-General B. N. Nunn	1987–90
Major-General W. E. Glenny	1990–91

*During these periods, the formation was known as 3rd Divisional Field Force Group.

Appendix 2
Photograpic Gallery of 3rd Australian Division Commanders
[*Courtesy* HQ 3rd Division]

A-1: Major General J. Monash
GOC 3rd Division, July 1916–June 1918

A-2: Major General J. Gellibrand
GOC 3rd Division, June 1918–April 1919
February 1921–February 1922

A-3: Major General G. J. Johnston
GOC 3rd Division, February 1922–April 1927

A-4: Major General H. E. Elliott
GOC 3rd Division, May 1927–March 1931

A-5: Major General T. A. Blamey
GOC 3rd Division, June 1931–May 1937

A-6: Major General E. A. Drake-Brockman
GOC 3rd Division, June 1937–February 1942

A-7: Major General S. Savige
GOC 3rd Division, January 1942–February 1944

A-8: Major General W. Bridgeford
GOC 3rd Division, July 1944–November 1947

A-9: Major General G. F. Wootten
GOC 3rd Division, November 1947–May 1950

A-10: Major General S. H. W. C. Porter
GOC 3rd Division, 2 May 1950–2 July 1953

A-11: Major General R. J. H. Risson
GOC 3rd Division, July 1953–June 1956

A-12: Major General H. H. Hammer
GOC 3rd Division, June 1956–July 1959

A-13: Major General N. W. Simpson
GOC 3rd Division, July 1959–November 1960

A-14: Major General R. R. Gordon
GOC 3rd Division, December 1960–November 1963

A-15: Major General N. A. Vickery
GOC 3rd Division, December 1963–November 1966

A-16: Major General S. M. McDonald
GOC 3rd Division, December 1966–January 1970

A-17: Major General K. D. Green
GOC 3rd Division, January 1970–April 1973

A-18: Major General J. M. McNeill
GOC 3rd Division, May 1973–February 1976

A-19: Brigadier W. H. Grant
Comd 3rd Division Field Force Group, February 1976–March 1977

A-20: Brigadier J. E. Barry
Comd 3rd Division Field Force Group, April 1977–January 1980
and Major General GOC 3rd Division April 1985–March 1987

A-21: Brigadier N. E. Bavington
Comd 3rd Division Field Force Group, January 1980–April 1981

A-22: Major General K. G. Cooke
Comd 3rd Division Field Force Group, July 1981–March 1984
GOC 3rd Division, April 1984–March 1985

A-23: Major General B. N. Nunn
GOC 3rd Division, March 1987–March 1990

A-24: Major General W. E. Glenny
GOC 3rd Division, March 1990–March 1991

Bibliography

Primary Sources: Manuscripts

Australian Archives, Canberra

A816	Department of Defence, Correspondence Files, 1935-57
A2653	Volumes of the Military Board Proceedings, 1905-76
A2657	Volumes of Papers of Historical Interest, 1909
A3688	Department of the Army, Correspondence Files, 1946-79
A5954	Shedden Collection, 1937-71
A6059	Department of the Army, Correspondence Files, 1956-64
A6835	CGS Outward Correspondence Files, 1963-74
A6922	Policy and Working Files of the Directorate of Staff Duties of Army Headquarters, 1939-73

Australian Archives, Melbourne

B197	Department of Defence, Correspondence Files, 1906-36
B420	3rd Division Headquarters Files
B1535	Department of Army, Correspondence Files, 1930-39
MP70	Headquarters 3rd District Base
MP367	Department of Defence, Correspondence Files, 1917-29
MP385	Headquarters Victorian District

Australian Defence Force Academy Library

Wilton Collection

Australian War Memorial

AWM1	Pre-Federation and Commonwealth Records
AWM4	War Diaries, World War I
AWM26	Operational Files
AWM27	Subject Classified Files
AWM28	Awards and Honours
AWM49	Army Records, 1919-42
AWM52	War Diaries, World War II
AWM54	Written Records, 1939-45
AWM62	Southern Command Records
AWM113	Records of the Military History Section (Army), 1940-62
AWM224	3rd Division Artillery Reports
1DRL136	Boulton Collection
1DRL545	Peters Collection
2DRL003	Rose Collection
2DRL0059	Atkinson Collection
2DRL185	Carson Collection
2DRL234	Schwinghammer Collection
3DRL1473	Gellibrand Collecton

3DRL2147	Robinson Collection
3DRL2316	Monash Collection
3DRL2529	Savige Collection
3DRL3561	Stevens Collection
3DRL6461	Greenlough Collection
3DRL6643	Blamey Collection
3DRL7137	Mowat Collection
3DRL7528	Baker Collection
MS1255	Peters, 'Raiding Days'
PR00266	Finney Collection
PR00446	Byrne Collection
PR00519	Hardie Collection
PR00594	Murray Collection
PR00618	Crowley Collection
PR91/002	Moten Collection
PR91/046	Bellingham Collection

Defence Central Registry

A6721 Department of Defence, Year and Single File Series, 1985–present

Victoria Barracks, Melbourne

MP1481 Southern Command Correspondence Files

Primary Sources: Published

Chauvel, H. G. *Report of the Inspector-General of the Australian Military Forces, 1921*. Melbourne: Government Printer, 1921.

——. *Report of the Inspector-General of the Australian Military Forces, 1922*. Melbourne: Government Printer, 1922.

——. *Report of the Inspector-General of the Australian Military Forces, 1923*. Melbourne: Government Printer, 1923.

——. *Report of the Inspector-General of the Australian Military Forces, 1930*. Melbourne: Government Printer, 1930.

Commonwealth Parliamentary Papers.

Department of Defence, *The Defence of Australia*. Canberra: AGPS, 1987.

Dibb, Paul. *Review of Australia's Defence Capabilities*. Canberra: AGPS, 1986.

Force Structure Review, 1991. Canberra, AGPS, 1991.

Hamilton, Ian. *Report on an Inspection of the Military Forces of the Commonwealth of Australia*. Melbourne: Government Printer, 1914.

Kirkpatrick, G. M. *Annual Report of the Inspector-General of the Military Forces*. Melbourne: Government Printer, 1913.

Report of the Committee of Inquiry into the Citizen Military Forces. Canberra: AGPS, 1974.

Savige, S. G. *Tactical and Administrative Doctrine for Jungle Warfare*. HQ II Australian Corps, 1945.

Secondary Sources: Published

Anon. *42nd Battalion, AIF*. Brisbane, 1938.

——. *Getting Through: The Unit War History of 3rd Australian Division Signals, 1939-1945*. Dandenong, Vic: Kenwalk Printing, 1987.

——. *The History of the Ninth Field Ambulance*. Sydney: John Sands, 1919.

——. *History, 10th Field Company, Engineers.* 1919.

——. *Short History of the 34th Battalion, A. I. F.* 34th Battalion A. I. F. Association.

——. *Souvenir, Being an Unofficial Résumé of the History of the 11th Australian Field Ambulance.* Eleventh Australian Field Ambulance, 1919.

——. *The Thirty-ninth: The History of the 39th Battalion Australian Imperial Force.* Melbourne: G. W. Green & Sons, 1934.

Austin, Victor. *To Kokoda and Beyond: The Story of the 39th Battalion, 1941-1943.* Carlton, Vic: Melbourne University Press, 1988.

Bean, C. E. W. *The Australian Imperial Force in France, 1916.* St Lucia: University of Queensland Press, 1982.

——. *The Australian Imperial Force in France, 1917.* Sydney: Angus & Robertson, 1943.

——. *The Australian Imperial Force in France During the Main German Offensive, 1918.* Sydney: Angus & Robertson, 1937.

Beaumont, Joan, 'Australia's War: Asia and the Pacific' in Joan Beaumont, ed., *Australia's War, 1939-45.* St Leonards: Allen and Unwin, 1996.

Benson, S. E. *The Story of the 42nd Australian Infantry Battalion.* Sydney: 42nd Australian Infantry Battalion, 1952.

Blair, Ron. *A Young Man's War: A History of the 37th/52nd Australian Infantry Battalion in World War Two.* Melbourne, 37th/52nd Australian Infantry Battalion Association, 1992.

Bourne, Jack and Jack Lavery. *A Purple Patch: A History of the 10th Field Company, Royal Australian Engineers (AIF) 1940-1945.* Melbourne: 10th Field Company, 1991.

Brahms, Vivian. *The Spirit of the Forty-Second: Narrative of the 42nd Battalion, 11th Infantry Brigade, 3rd Division, Australian Imperial Forces during the Great War, 1914-1918.* Brisbane: Smith & Paterson, 1938.

Bridger, T. D. *With the 27th Battery in France.* London: St Clements Press, 1919.

Broadhurst, William. *The Young Soldiers: The Battle of the Mivo Crossing, the 1945 Experience of the 15th Battalion Mortar Platoon*, np, nd.

Budden, F. M. *The Chocos: The Story of the Militia Infantry Battalions in the South West Pacific Area, 1941-1945.* Summer Hill, NSW: F. M. Budden, 1987.

Butler, A. G. *The Australian Army Medical Service in the War of 1914-18*, vol. III, *Special Problems and Services.* Canberra: Australian War Memorial, 1943.

Charlott, Rupert, ed. *The Unofficial History of the 29th/46th Australian Infantry Battalion, AIF, September 1939 – September 1945.* Melbourne: 1952.

Charlton, Peter. *The Unnecessary War: Island Campaigns of the South-West Pacific, 1944-45.* South Melbourne: MacMillan, 1983.

Christensen, George, ed. *That's the Way it Was: The History of the 24th Australian Infantry Battalion (A.I.F.), 1939-1945.* Hawthorn, Vic: 24th Battalion Association, 1982.

Colliver, E. J. and B. H. Richardson. *The Forty-Third: The Story and Official History of the 43rd Battalion, AIF.* Adelaide: Rigby, 1920.

Cooke, K. G. 'One Army' in Peter Dennis and Jeffrey Grey, eds. *The Second Fifty Years: The Australian Army , 1947-1997.* Canberra: Australian Defence Force Academy, 1997.

Corfield, Robin S. *Hold Hard Cobbers, The Story of the 57th and 60th and 57/60th Australian Infantry Battalions, 1912-1990*, vol. 2, *1930-1990.* Glenhuntly, Vic: 57/60th Battalion Association, 1991.

Cutlack, F. M. ed. *The War Letters of General Monash.* Sydney: Angus & Robertson, 1935.

Cuttriss, G. P. *'Over the Top' With the Third Australian Division.* London: Charles H. Kelly, 1918.

Dawes, J. N. I. and L. L. Robson. *Citizen to Soldier: Australia Before the Great War, Recollections of Members of the First AIF.* Carlton, Vic: Melbourne University Press, 1977.

Dennis, Peter, et al. *The Oxford Companion to Australian Military History*. Melbourne: Oxford University Press, 1997.

Dexter, David. *The New Guinea Offensives*. Canberra: Australian War Memorial, 1961.

Edwards, John. *Never a Backward Step: A History of the First 33rd Battalion, AIF*. Grafton: Bettong Books, 1996.

Edwards Peter. *Crises and Commitments: The Politics and Diplomacy of Australia's Involvement in South East Asian Conflicts, 1948-1965*. North Sydney: Allen & Unwin, 1992.

Fairey, Eric. *The Story and Official History of the 38th Battalion, AIF*. Bendigo, Vic: Cambridge Press, 1920.

Festburg, Alfred, N. *The Lineage of the Australian Army*. Melbourne: Allara Publishing, 1972.

Gailey, Harry A. *Bougainville 1943-1945: The Forgotten Campaign*. Lexington, KY: The University of Kentucky, 1991.

Gammage, Bill. *The Broken Years: Australian Soldiers in the Great War*. Canberra: Australian National University Press, 1974.

Green, F. C. *The Fortieth: A Record of the 40th Battalion, AIF*. Hobart: Government Printer, 1922.

Grey, Jeffrey. *A Military History of Australia*. Melbourne: Cambridge University Press, 1999.

Hamill, Ian. *The Strategic Illusion: The Singapore Strategy and The Defence of Australia and New Zealand, 1919-1942*. Singapore: Singapore University Press, 1981.

Hetherington, John. *Blamey: The Biography of Field-Marshal Sir Thomas Blamey*. Melbourne: F. W. Chesire, 1954.

Horner, David. 'Staff Corps Versus Militia: The Australian Experience in World War II', *Defence Force Journal*, no. 26 (Jan/Feb 1981), pp. 13-25.

——. *Blamey: The Commander in Chief*. St Leonards: Allen and Unwin, 1998.

Jones, A. Les. *The Mail Must Get Through*, np, 1983.

Kagan, Donald. *On the Origins of War and the Preservation of Peace*. New York: Anchor Books, 1995.

Kaighin, Thomas R. L. *A Personal Account of World War I*. Rodd Johnson, transcriber. n.p.: Rodd Johnson, 2000.

Keatinge, M. B. B. *War Book of the Third Pioneer Battalion*. Melbourne: The Specialty Press, 1934.

Leckie, Neil. *Bushmens Rifles: A History of the 22nd Battalion, The Royal Victoria Regiment, 1966-1975*. Mulgrave, VIC: The Ex 22nd Bn RVR Association, 1999.

Long, Gavin. *The Final Campaigns*. Canberra: Australian War Memorial, 1963.

Longmore, C. *'Eggs-A-Cook' The Story of the Forty-Fourth, War as the Digger Fought It*. Perth: The Colortype Press, 1920.

Mathews, Russell. *Militia Battalion at War: The History of the 58th/59th Australian Infantry Battalion in the Second World War*. Sydney: 58th/59th Battalion Association, 1961.

McCarthy, Dudley. *South-West Pacific Area – First Year, Kokoda to Wau*. Canberra: Australian War Memorial, 1962.

McCarthy, John. *Australia and Imperial Defence 1918-1939: A Study in Air and Sea Power*. St Lucia: University of Queensland Press, 1976.

McNicol, H. G. *The Thirty-Seventh: A History of the Thirty-Seventh Battalion AIF*. Melbourne: Modern Printing Co., 1936.

Medcalf, Peter. *War in the Shadows: Bougainville, 1944-45*. Sydney and Canberra: Collins and Australian War Memorial, 1986.

Monash, John. *The Australian Victories in France in 1918*. London: Hutchinson & Co., 1920.

Nicholls, Bob. *The Colonial Volunteers: The Defence Forces of the Australian Colonies, 1836-1901*. North Sydney: Allen & Unwin, 1988.

Pedersen, P. A. *Monash as Military Commander*. Carlton, Vic: Melbourne University Press, 1985.

Pipeclay, The Magazine of the Melbourne Naval & Military Club, 1991.

Prior, Robin and Trevor Wilson. *Command on the Western Front: The Military Career of Sir Henry Rawlinson, 1914-1918*. Oxford: Blackwell Publishers, 1992.

Reddrop, Mary, ed. *Jim's Story: With the 37th Battalion – AIF*. Melbourne: Spectrum Publications, 1982.

Robertson, John. *Australia at War: 1939-1945*. Melbourne: William Heinemann, 1981.

Ross, A. R. ed. *The Magazine of the 17th Australian Infantry Brigade, 1939-1944*, np, 1944.

Russell, W. B. *There Goes a Man: The Biography of Sir Stanley G. Savige*. Melbourne, Longmans, 1959.

Sadler, Peter S. *The Paladin: A Life of Major-General Sir John Gellibrand*. Melbourne: Oxford University Press, 2000.

Schacht, Tod. *My War on Bougainville*. Loftus: Australian Military History Publications, 1999.

Scott, Ernest. *Australia During the War*. Sydney: Angus & Robertson, 1937.

Serle, Geoffrey. *John Monash: A Biography*. Carlton, Vic: Melbourne University Press, 1982.

Taylor, Frank. *Mad Mick's Mob: A History of the 15th Australian Field Company, RAE (AIF)*. Mordialloc, Vic: 1995.

Wilcox, Craig. *For Hearths and Homes: Citizen Soldiering in Australia, 1854-1945*. St Leonards: Allen & Unwin, 1998.

Secondary Sources: Unpublished

Anon, 'History of the Royal Australian Engineers, 3rd Division CMF'.

Burns, Jock. 'The Formation of the Third Division, AIF in 1916 and Service in World War I'.

Keating, G. M. 'The Right Man for the Job: An Assessment of Lieutenant-General S. G. Savige as a Senior Commander', Thesis submitted for the Degree of Bachelor of Arts (Honours), School of History, University College, The University of New South Wales, Australian Defence Force Academy (1995).

McCarthy, D. S. 'The Once and Future Army: An Organisational, Political and Social History of the Citizen Military Forces, 1947-1974', Thesis submitted for the degree of Doctor of Philosophy in the School of History, University College, Australian Defence Force Academy, University of New South Wales, 1997.

Murphy, J. E. 'History of the Post-War Army'.

Rae, Ian D. 'History of the Australian Army, Part 1: Early History'.

Index